T0258125

Handbook of Digital Communication

Handbook of Digital Communication

Edited by **Bernhard Ekman**

New York

Published by NY Research Press,
23 West, 55th Street, Suite 816,
New York, NY 10019, USA
www.nyresearchpress.com

Handbook of Digital Communication
Edited by Bernhard Ekman

International Standard Book Number: 978-1-63238-233-7 (Hardback)

Printed in the United States of America.

Contents

Permissions

Preface

This book provides state-of-the-art information regarding digital communications. Everyone should have a digital strategy since all marketing is digital these days. Everything is going mobile. The current talk in the digital community is that "the world has never been more social" and digital communication is considered as the key facilitator of this fact. Digital information tends to be much more defiant to disseminate and decipher errors than information symbolized in an analog medium. This accounts for the clarity of digitally-encoded compact audio disks, telephone connections and a lot of enthusiasm for digital communications technology in the engineering community. With a modern and descriptive presentation approach regarding the field of digital communication, this book explores modernized digital communication methodologies. The aim of this book is to update and enhance the knowledge of the reader regarding the dynamically transforming field of digital communication.

All of the data presented henceforth, was collaborated in the wake of recent advancements in the field. The aim of this book is to present the diversified developments from across the globe in a comprehensible manner. The opinions expressed in each chapter belong solely to the contributing authors. Their interpretations of the topics are the integral part of this book, which I have carefully compiled for a better understanding of the readers.

At the end, I would like to thank all those who dedicated their time and efforts for the successful completion of this book. I also wish to convey my gratitude towards my friends and family who supported me at every step.

Editor

Wireless Communication in Tunnels

Jose-Maria Molina-Garcia-Pardo[1], Martine Lienard[2] and Pierre Degauque[2]
[1]Universidad Politécnica de Cartagena
[2]University of Lille, IEMN
[1]Spain
[2]France

1. Introduction

A great deal of work is being done to optimize the performances of digital wireless communication systems, and most of the effort is focused on the urban environment where technologies evolve surprisingly fast. Deployment of new emerging technologies, all based on digital communications, first requires the knowledge of the physical layer in order to develop efficient antenna design and communication algorithms. Tunnels, including rail, road and pedestrian tunnels, even if they do not represent a wide coverage zone, must be taken into account in the network architecture, the tunnel either being considered part of a neighboring cell or as a cell itself. In order to cover the tunnel, two solutions have traditionally been proposed: the so-called "natural propagation" using antennas of small size, and leaky coaxial cables. However, the implementation of radiating cables is expensive, at least in long tunnels, because the diameter of the cable must be large enough to avoid a prohibitive attenuation in the 1-5 GHz band, which will be considered in this chapter. We will thus focus our attention on a link based on natural propagation.

Whatever the application, preliminary knowledge of the propagation phenomena is required. The first section of this chapter is thus devoted to a presentation of theoretical models, while in the second section the main narrow band and wideband double-directional channel characteristics, determined from numerous measurement campaigns, will be presented and interpreted. Since Multiple-Input Multiple-Output (MIMO) techniques may strongly improve the spectral efficiency and/or decrease the error rate, keeping the transmitting power and the bandwidth constant; the last two sections will describe their performances in tunnels. Indeed, one can expect the degree of diversity of the channel to be, by far, quite different from its average value in an indoor environment due to the guiding structure of the tunnel. Applying the propagation model to MIMO allows the outlining of the main parameters playing an important role on the ergodic channel capacity, and introducing the so-called modal diversity. From measured channel matrices, predicted capacity is given for various tunnel and array configurations. The last paragraph of this chapter treats the robustness, in terms of error rate, of different MIMO schemes.

2. Modeling the propagation channel

Many approaches have been developed to theoretically study electromagnetic wave propagation inside a tunnel, the most well known being those based either on the ray theory or on the modal theory (Mahmoud, 1988; Dudley et al., 2007). The transmitting frequency range must be chosen such that the attenuation per unit length is not prohibitive. To fulfill this requirement, the tunnel must behave as an oversized waveguide. Consequently, the wavelength must be much smaller than the transverse dimensions of the tunnel, which leads to transmitting frequencies greater than few hundred MHz in usual road or train tunnels. The objective of this section is to conduct an overview of the techniques to treat tunnels of simple geometry, such as rectangular or circular straight tunnels, by using either the ray theory or the modal theory. Studying wave propagation along such structures will allow simple explanation and interpretation of the experimental results obtained in real tunnels, even of more complicated shapes. Theoretical approaches to treat tunnels of an arbitrary cross-section and/or presenting a series of curves will also be briefly presented.

2.1 Propagation in a straight rectangular tunnel

Let us consider a rectangular tunnel along the z-axis, as shown in Fig. 1, the width and the height of the tunnel being equal to a and b, respectively. The coordinate origin is in the centre of the cross-section, at $z = 0$, which defines the excitation plane. The walls are either characterized by their complex permittivity ε_r^* or by an equivalent conductivity σ and real permittivity ε_r.

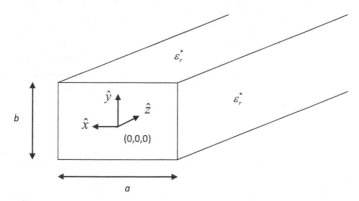

Fig. 1. Geometry of the rectangular tunnel.

2.1.1 Ray theory

Ray theory combined with image theory leads to a set of virtual transmitting (Tx) antennas. If the tunnel is of rectangular cross-section, the determination of the location of these virtual antennas is straightforward, and it is independent of the location of the receiving point. The total field is obtained by summing the contribution of all rays connecting the Tx images and the receiving point (Rx), whilst considering the reflection coefficients on the tunnel walls. However, even by assuming ray propagation, the summation of the contribution of the rays at the Rx point must take into account the vector nature of the electric field. Before each reflection on a wall, the electric field vector must be expressed as the sum of two components: one perpendicular to the incidence plane E_{perp} and one parallel to this plane E_{para}. To each of these components, reflection coefficients R_{TM} and R_{TE} are respectively applied, mathematical expressions for which can be found in any book treating electromagnetic wave propagation (Wait, 1962; Dudley, 1994). After each reflection, one can thus obtain the new orientation of the electric field vector. The same approach is successively applied by following the rays and by finding the successive incidence planes. Nevertheless, it has been shown (Lienard et al., 1997) that if the distance between Tx and Rx becomes greater than three times the largest dimension of the tunnel cross-section, the waves remain nearly linearly polarized. In this case, the vector summation of the electric field radiated by an antenna and its images becomes a simple scalar summation as in:

$$E(x,y,z) = \sum_m \sum_n (R_{TM})^m (R_{TE})^n E_d(S_{mn}) \tag{1}$$

where $E_d(S_{mn})$ is the electric field radiated in free space by the image source S_{mn} and corresponding to rays having m reflections on the walls perpendicular to the Tx dipole axis and n reflections on the walls parallel to the dipole axis (Mahmoud & Wait, 1974). In the following examples the excitation by an electric dipole will be considered, but this is not a strong restriction in the ray approach since other kinds of antennas can be treated by introducing their free space radiation pattern into the model, i.e. by weighting the rays in a given direction by a factor proportional to the antenna gain in this direction.

Lastly, it must also be emphasized that the reflection coefficients on the walls tend to 1 if the angle of incidence on the reflecting plane tends to 90°. This means that, at large distances, only rays impinging the tunnel walls with a grazing angle of incidence play a leading part in the received power, thus the number of rays that fulfill this condition is important. Typically, to predict the total electric field in standard tunnels and at distances of a few hundred meters, 20 to 30 rays are needed. One should note that if a base station is located outside the tunnel, and if a mobile moves inside the tunnel, the ray theory can still be applied by taking the diffraction in the aperture plane of the tunnel into account (Mariage et al., 1994).

2.1.2 Modal theory

The natural modes propagating inside the tunnel are hybrid modes EH_{mn}, the three components of the electric and magnetic fields that are present (Mahmoud, 2010; Dudley et al., 2007). Any electric field component $E(x, y, z)$ can be expressed as a sum of the modal components:

$$E(x,y,z) = \sum_m \sum_n A_{mn}(0)\, e_{mn}(x,y)\, e^{-\gamma_{mn} z} \tag{2}$$

In this expression, $A_{mn}(0)$ is the complex amplitude of the mode in the excitation plane, e_{mn} is the normalized modal eigenfunction, and γ_{mn} is the complex propagation constant, often written as $\gamma_{mn} = \alpha_{mn} + j\beta_{mn}$.

It is interesting to introduce the weight of the modes $A_{mn}(z)$ at any abscissa z by stating:

$$E(x,y,z) = \sum_m \sum_n A_{mn}(z)\, e_{mn}(x,y) \quad \text{where} \quad A_{mn}(z) = A_{mn}(0)\, e^{-\gamma_{mn} z} \tag{3}$$

The analytical expressions of the modal eigenfunctions are usually obtained by writing the boundary conditions on the internal surface of the guiding structure. However, in the case of lossy dielectric walls, as in the case of a tunnel, approximations are needed and they are detailed in (Emslie et al., 1975; Laakman & Steier, 1976). We have previously outlined that rays remain polarized if the distance between Tx and Rx is larger than a few times the transverse dimensions of the tunnel. In the modal theory, we have the same kind of approximation. If the tunnel is excited by a vertical (y-directed) dipole, the hybrid modes EH_{mn}^y are such that the vertical electric field is dominant. For an x-directed dipole, the modes are denoted EH_{mn}^x. The expressions of the modal functions $e_{mn}^V(x,y)$ for the y-polarized modes, and $e_{mn}^H(x,y)$ for the x-polarized modes can be found in (Mahmoud, 2010; Dudley et al., 2007). The solution of the modal equation leads to expressions for the phase and attenuation constants for the y-polarized modes:

$$\alpha_{mn} = \frac{2}{a}\left(\frac{m\lambda}{2a}\right)^2 \operatorname{Re}\left[\frac{1}{\sqrt{\varepsilon_r^* - 1}}\right] + \frac{2}{b}\left(\frac{n\lambda}{2b}\right)^2 \operatorname{Re}\left[\frac{\varepsilon_r^*}{\sqrt{\varepsilon_r^* - 1}}\right] \text{ and } \beta_{mn} = \frac{2\pi}{\lambda}\left[1 - \frac{1}{2}\left(\frac{m\lambda}{2a}\right)^2 - \frac{1}{2}\left(\frac{n\lambda}{2b}\right)^2\right] \quad (4)$$

From (4), we see that the attenuation is inversely proportional to the waveguide dimension cubed and the frequency squared. It must be stressed that, given the finite conductivity of the tunnel walls, the modes are not precisely orthogonal. Nevertheless, numerical applications indicate that, when considering the first 60 modes with orders m≤ 11 and n≤ 7, the modes can be considered as practically orthogonal (Lienard et al., 2006; Molina-Garcia-Pardo et al., 2008a). As an example, let us consider a tunnel whose width and height are equal to 4.5 m and 4 m, respectively, and whose walls are characterized by an equivalent conductivity σ = 10⁻² S/m and a relative real permittivity ε_r = 10. Table 1 gives the attenuation, expressed in dB/km, of the first $EH_{m,n}$ hybrid modes and for two frequencies 2.4 GHz and 10 GHz.

	n=1	n=2	n=3		n=1	n=2	n=3
m=1	4.6 dB	17 dB	38 dB	m=1	0.2 dB	0.8 dB	1.8 dB
m=2	6 dB	18 dB	40 dB	m=2	0.3 dB	0.9 dB	1.9 dB
m=3	8 dB	21 dB	42 dB	m=3	0.4 dB	1 dB	2 dB
f = 2.4 GHz.				f = 10 GHz			

Table 1. Attenuation along 1 km of the various $EH_{m,n}$ hybrid modes

At 10 GHz, the fundamental mode exhibits a negligible attenuation, less than 0.2 dB/km, while at 2.4 GHz it is in the order of 5 dB/km. Two other points must be outlined. First, if we consider, for example, a frequency of 2.4 GHz, and a distance of 1 km, one can expect that only 2 or 3 modes will play a leading part in the total received signal, while at 10 GHz, a large number of modes will still be present, the attenuation constant being rather low. This leads to the concept of the number N_a of "active modes" significantly contributing to the total power at the receiver, and which will be extensively used in 4.1 to predict the capacity of multiantenna systems such as MIMO. To simply show the importance of introducing these active modes, let us recall that the phase velocity of the waves differs from one mode to another. Interference between modes occurs, giving rise to fluctuations in the signal both along the z-axis and in the transverse plane of the tunnel. One can expect that at 1 km, the fluctuations at 10 GHz will be more significant and rapid than at 2.4 GHz, taking the large number of "active modes" into account. Similarly, for a given frequency N_a continuously decreases with the distance between Tx and Rx. Variations in the field components will thus be more pronounced in the vicinity of Tx. This phenomenon will have a strong impact on the correlation between array elements used in MIMO systems.

For a vertical transmitting elementary dipole situated at (x_{tx}, y_{tx}), the total electric field at the receiving point (x, y, z) can be determined from (Molina-Garcia-Pardo et al., 2008c):

$$E(x,y,z) = \sum_m \sum_n e_{m,n}^V (x_{tx}, y_{tx}) \, e_{m,n}^V (x,y) \, e^{-\gamma_{m,n}z} \qquad (5)$$

To treat the more general case of a radiating structure presenting a radiation pattern which is assumed to be known in free space, the easiest solution to determine the weight of the modes in the Rx plane is to proceed in two steps. First, the E field in the tunnel is calculated numerically using, for example, the ray theory, and then the weight $A_{m,n}(z)$ of any mode m,n for this abscissa can be obtained by projecting the electric field on the basis of $e_{mn}^V(x,y)$ (Molina-Garcia-Pardo et al., 2008a). This leads to:

$$A_{mn}(z) = \int_{-a/2}^{a/2} \int_{-b/2}^{b/2} E(x,y,z) \cdot e_{mn}^V(x,y) \, dxdy \qquad (6)$$

In the following numerical application a 8 m-wide, 4.5 m-high tunnel is considered, the electrical parameters of the walls being $\sigma = 10^{-2}$ S/m and $\varepsilon_r = 5$. The radiating element is a vertical elementary dipole situated 50 cm from the ceiling, at 1/4 of the tunnel width, and the transmission frequency is 900 MHz. This configuration could correspond to a practical location of a base station antenna in a real tunnel. The six modes with the highest energies at a distance of 300 m and 600 m are provided in Table 2.

Modes	2,1	1,1	3,1	5,1	6,1	2,2
300 m	0 dB	-2 dB	-3 dB	-4 dB	-4 dB	-5 dB
600 m	0 dB	-2 dB	-4 dB	- 9 dB	-9 dB	-15 dB

Table 2. Relative weights of the modes at 300 m and 600 m.

Due the non-centered position of the Tx dipole, the most energetic mode is mode 2,1, although mode 1,1 is the lowest attenuated. The other columns in Table 2 show the relative weight of the other modes, normalized for each distance in terms of the highest mode weight, i.e. mode 2,1. At 300 m, numerous other modes are still significantly contributing to the total field, since the 6th mode only presents a relative attenuation of 5 dB in relation to the strongest mode. On the other hand, at 600 m only a few higher-order modes remain. The application of such an approach for predicting the performance of MIMO systems will be used in 4.1 and 4.2.

2.2 Propagation in a straight circular tunnel

Even if a perfectly circular tunnel is less usual, it is interesting to outline some specific features related to the modes and polarization of the waves propagating in such a guiding structure. A detailed analysis (Dudley et al., 2007) shows that an elementary electric dipole produces a large set of modes, the possible modes being TE_{0m}, TM_{0m} and the hybrid modes EH_{nm} and HE_{nm}.

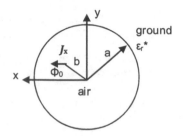

Fig. 2. Configuration of a cylindrical tunnel and location of a transmitting antenna.

An important feature of the propagation phenomena for MIMO communication systems based on polarization diversity is the cross polarization discrimination factor, XPD, defined as the ratio of the co-polarized to the cross-polarized average received power. Indeed, as will be outlined in 3.1.2, XPD quantifies the separation between two transmission channels that use different polarization orientations. In a circular tunnel and assuming a x-oriented dipole Jx at $(b, \phi_0, 0)$, it can be shown (Dudley et al., 2007) that the cross-polar fields at an observation point (ρ, ϕ, z) vanish at $\phi_0 = 0, \pi / 2$ but are maximum at $\phi_0 = \pi / 4$. In this last case, if the observation point is at the same circumferential location as the source point, XPD becomes equal to 1. The polarization of the waves in a circular tunnel is thus quite different than in a rectangular tunnel, where, at large distances, the co-polarized field component is always dominant.

In practice, the shape of a tunnel is often neither perfectly circular nor rectangular. Consequently, numerous measurement campaigns have been carried out in different tunnel configurations and the results, compared to the theoretical approach based on simplified shapes of tunnel cross-sections, are presented in 3.2. However, before presenting narrow band and wideband channel characteristics in a real tunnel, the various methods for numerically treating propagation in tunnels of arbitrary shape will be briefly described.

2.3 Propagation in a tunnel of arbitrary shape

If we first consider a bent tunnel of rectangular cross-section whose radius of curvature is much larger than the transverse dimensions of its cross-section, approximate solutions of the modal propagation constants based on Airy function representation of the fields have been obtained (Mahmoud, 2010). For an arc-shaped tunnel, the deviation of the attenuation and the phase velocity of the dominant modes from those in a perfectly rectangular tunnel are treated in (Mahmoud, 2008) by applying a perturbation theory. Lastly, let us mention that for treating the propagation in tunnels of arbitrary shapes, various approaches have recently been proposed, despite the fact that they are more complicated to implement and that the computation time may become prohibitive for long-range communication. Ray launching techniques and ray-tube tracing methods are described in (Didascalou et al., 2001) and in (Wang & Yang, 2006), the masking effect of vehicles or trains being treated by introducing additional reflection/diffraction on the obstacle. A resolution of the full wave Maxwell equations in the time domain through a high order vector finite element discretization is proposed in (Arshad et al., 2008), while solutions based on the parabolic equation and spectral modeling are detailed in (Popov & Zhu, 2000). We will not describe in detail all these methods, since the objective of this chapter is to present the general behavior of the propagation in tunnels, rather than to emphasize solutions to specific problems.

3. Narrow band and wideband channel characterization

As previously outlined, a number of propagation measurements in tunnels have been taken over the last 20 years, and it is not within the scope of this chapter to make an extensive overview of what has been done and published. Let us simply mention that the results cover a wide area of environment and applications, starting from mine galleries (Lienard & Degauque, 2000a; Zhang et al., 2001; Boutin et al. 2008), road and railway tunnels (Lienard et al., 2003; Lienard et al., 2004), pedestrian tunnels (Molina-Garcia-Pardo et al., 2004), etc. We have thus preferred to select one or two scenarios throughout this chapter in order to clearly identify the main features of the propagation phenomena and their impact on the optimization of the transmission scheme.

3.1 Narrow band channel characteristics

We first consider the case of a rectangular tunnel for applying the simple approach described in the previous sections and for giving an example of path loss versus the distance between Tx and Rx. Then, results of the experiments carried out in an arched tunnel will be presented and we will study the possibility of interpreting the measurements from a simple propagation model, i.e. by means of an equivalent rectangular tunnel.

3.1.1 Path loss determined from a propagation model in a rectangular tunnel

Let us consider a wide rectangular, 13 m wide and 8 m high, corresponding to the transverse dimensions of a high-speed train tunnel. Curve (a) in Fig. 3 represents the variation of the field amplitude, expressed in dB and referred to an arbitrary value, versus the distance d, which is determined from the theoretical approach based on the ray theory. The transmitting frequency is 2.1 GHz and the Tx antenna is supposed to be a half-wave dipole situated at a height of 2 m and at a distance of 2 m along a vertical wall.

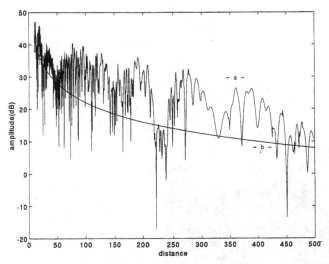

Fig. 3. a) Amplitude of the received signal, referred to an arbitrary value, determined from the ray theory in a rectangular tunnel. b) Amplitude of the signal for a free space condition (Lienard & Degauque, 1998).

The walls have a conductivity of 10^{-2} S/m and a relative permittivity of 10. As a comparison, curve (b) corresponds to the case of free space propagation. Along the first 50 m the field amplitude decreases rapidly, in the order of 5 dB/100 m, and fast fluctuations can be observed. They are due to the interference between the numerous paths relating Tx and Rx or, from the modal theory point of view, to interference between modes. Beyond this distance d_1 of 50 m, we note a change in the slope of the path loss which becomes equal to 1 dB/100 m. A two-slope model is thus well suited for such a tunnel and is also detailed in (Marti Pallares et al., 2001; Molina-Garcia-Pardo et al., 2003). In general, the abscissa of the break point d_1 depends on the tunnel excitation conditions and hence, on the position of the Tx antenna in the transverse plane, and on its radiation pattern. For a longer transmission range, a three/four-slope model has also been proposed (Hrovat et al., 2010).

After having determined the regression line in each of the two intervals and subtracted the effect of the average attenuation, one can study the fading statistics. Usually, as in an urban environment, the fading is divided into large scale and small scale fading, by considering a running mean on a few tens of wavelengths (Rappaport, 1996). However, in a straight tunnel, making this distinction between fading cannot be related to any physical phenomena. Furthermore, as it appears from curve 3a, the fading width and occurrence depend on the distance. An analysis of the fading characteristics is given in (Lienard and Degauque, 1998). Let us mention that the masking effect due to traffic in road tunnels or to trains in railway tunnels are detailed in (Yamaguchi et al., 1989; Lienard et al., 2000b; Chen et al., 1996, 2004).

3.1.2 Experimental approach to propagation in an arched tunnel and the concept of an equivalent rectangular tunnel

Numerous measurements have been carried out in the tunnel shown in Fig. 4a and 4b. The straight tunnel, 3 km-long, was closed to traffic during the experiments. The walls are made of large blocks of smooth stones. It is difficult to estimate the roughness accurately, but it is in the order of a few millimeters. In a first series of experiments, the transmitting power was 34 dBm

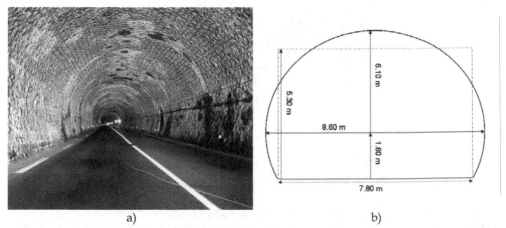

a)   b)

Fig. 4. a) Photo of the tunnel where measurements took place; b) Cross-section of the tunnel.

and the Tx and Rx half-wave dipole antennas were at the same height (2 m) and centered in the tunnel. Curves in Fig. 5 show the variation in the received power at 510 MHz, versus the axial distance d between Tx and Rx, the Tx and Rx antennas being both horizontally (HH) or vertically (VV) polarized (Molina-Garcia-Pardo et al., 2008b).

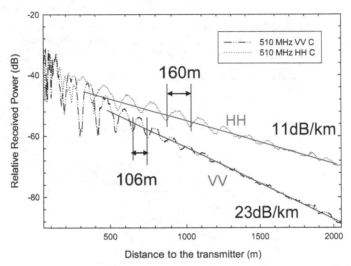

Fig. 5. Received power (dBm) in the arched tunnel for two polarizations at 510 MHz (Molina-Garcia-Pardo et al., 2008b).

We first note that the slope of the path loss is much higher for the VV polarization than for HH. Furthermore, if d becomes greater than 500 m, the spatial fading becomes periodic, but the periodicity is not the same for HH (160 m) and for VV (106 m). Since such results, showing a lack of cylindrical symmetry, cannot be interpreted with a simple model based on a cylindrical structure, we have tried to find the transverse dimensions of a rectangular tunnel, equivalent to the actual arched tunnel, which minimize the difference between the theoretical and experimental values. The choice of a rectangular cross-section whose surface is nearly equal to the surface of the actual tunnel seemed to be relevant. It appears that the best results were obtained for a rectangle 7.8 m wide and 5.3 m high, as shown in Fig. 4 b, the electrical characteristics of the walls being $\sigma = 10^{-2}$ S/m and $\varepsilon_r = 5$. To be useful, this equivalent rectangle must still be valid for other frequencies or configurations. As an example, let us now consider a frequency of 900 MHz, a vertical polarization (VV), both Tx and Rx vertical antennas being either centered in the tunnel (position noted C) or not centered (NC), i.e. situated at ¼ of the tunnel width (Molina-Garcia-Pardo et al., 2008c). Curves in Fig. 6 show the relative received power in dB (normalized to an arbitrary value) at 900 MHz for the two positions of the antennas ("C" and "NC") either measured or determined from the modal theory. In the theoretical modeling, only the first two dominant modes have been taken into account (EH_{11} and EH_{13} for "C", EH_{11} and EH_{12} for "NC").

A rather good agreement is obtained at large distance from Tx, where high order modes are strongly attenuated. The last point which can be checked before using the concept of equivalent rectangular tunnel, deals with the polarization of the waves at a large distance from Tx, cross-polar components appearing for critical positions of the Tx and Rx antennas in a circular tunnel, while in a rectangular tunnel the waves always remain polarized. Other measurements were made by placing the Tx and Rx antennas in the transverse plane at the critical location where XPD=1 in a circular tunnel ($\phi_0 = \pi / 4$ in Fig. 2). As detailed in Molina-Garcia-Pardo et al. (2008c), the experiments show that the waves remain polarized. The simple model of the propagation in an equivalent rectangular tunnel seems thus quite suitable to predict and/or justify the performances of communication systems using, for example, MIMO techniques, as described in 4.1 and 4.2.

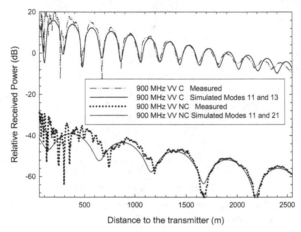

Fig. 6. Relative received power (dB) at 900 MHz for two configurations either measured or determined from the modal theory taking only two modes into account (Molina-Garcia-Pardo et al., 2008c).

3.2 Wideband and double directional channel characteristics

For optimizing the transmission scheme and predicting the performance of the communication link, the channel is characterized in the frequency domain by the coherence bandwidth B_c defined as the bandwidth over which the frequency correlation function is above a given threshold (Rappaport, 1996; Molisch, 2005), and in time domain by the delay spread D_S determined from the power delay profile (PDP). In the new generation of transmission schemes as in MIMO, double directional channel characteristics, such as the direction of arrival (DOA) and the direction of departure (DOD) of the rays are interesting to analyze since they have a strong impact on the optimization of the antenna arrays. We will successively present these characteristics established either from a propagation model or from measurements.

3.2.1 Characteristics determined from a propagation model in a rectangular tunnel

It has been previously outlined that, at a short distance from Tx, numerous modes or a large number of rays contribute to the total field. At a given distance of Tx, the number of active modes increases with the transverse dimensions of the tunnel, the attenuation of the EH_{mn} modes being a decreasing function of the width or of the height as shown in (4). If the tunnel width increases, interference between modes or rays will give a more rapidly fluctuating field with frequency, and thus a decrease in the coherence bandwidth B_c. As an example, for a frequency of 2.1 GHz in a tunnel 8 m high, B_C decreases from 40 MHz to 15 MHz when the tunnel width increases from 5 m to 20 m. The channel impulse response (CIR) can be established from the channel transfer function calculated in the frequency domain by applying a Hamming window, for example, and an inverse Fourier transform. In Fig. 7, for a tunnel 13 m wide and 8 m high and a bandwidth of 700 MHz, the theoretical amplitude of the received signal in a delay-distance representation has been plotted, in reference to an arbitrary level and using a color scale in dB. At a given distance, the successive packets of pulses, associated with reflections on the walls, clearly appear. We also see that the excess delay is a decreasing function of distance.

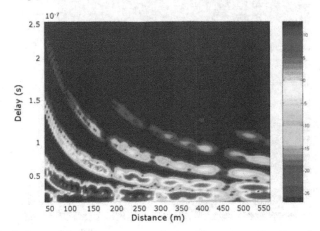

Fig. 7. Theoretical channel impulse response in a wide rectangular tunnel.

At first glance, one could conclude that the excess delay is a decreasing function of distance. However, a more interesting parameter is the delay spread, D_S, defined as the second-order moment of the CIR (Molisch, 2005). It is calculated at each distance d by normalizing each CIR to its peak value. For a threshold level of -25 dB, D_S varies in this wide tunnel, between 5 and 28 ns, but the variation is randomly distributed when 50 m < d < 500 m. From the theoretical complex CIRs, the DOA/DOD of the rays can be calculated by using high resolution algorithms such as SAGE or MUSIC (Therrien, 1992). At a distance d of 50 m, the angular spectrum density determined from MUSIC is plotted in Fig. 8. It represents the diagram (delay, DOA), each point being weighted by the relative amplitude of the received signal, expressed in dB above an arbitrary level, and conveyed in a color scale. The DOA is referred to the tunnel axis.

At a distance of 50 m, the rays arriving at an angle greater than 50° are strongly attenuated. A parametric study shows that the angular spread is a rapidly decreasing function of the distance d. This can be easily explained by the fact that rays playing a dominant role at large distances, impinging the tunnel walls with a grazing angle of incidence.

3.2.2 Characteristics determined from measurements in an arched tunnel

Measurement campaigns in the 2.8-5 GHz band have been carried out in the arched tunnel described in 3.1.2. The channel sounder was based on a vector network analyser (VNA) and on two virtual linear arrays. More details on the measurement procedure can be found in

(Molina-Garcia-Pardo et al., 2009c, 2009d). The mean delay spreads are given in Table 3, calculated in 5 successive zones, from 100 m to 500 m from Tx. As shown in Table 3, the delay spread in this range of distance remains nearly constant, in the order of a few ns, whatever the distance d. This result is strongly related to the DOA/DOD of the rays. The variation in their angular spread A_s is plotted in Fig. 9. At 50 m, A_s is equal to about 12° and then decreases with distance (Garcia-Pardo et al., 2011).

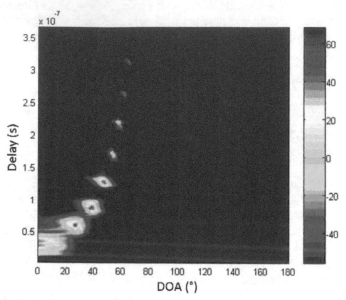

Fig. 8. Theoritical angular power spectrum of the DOA in the plane (delay, direction of arrival), in a wide tunnel and at 50 m from the transmitter.

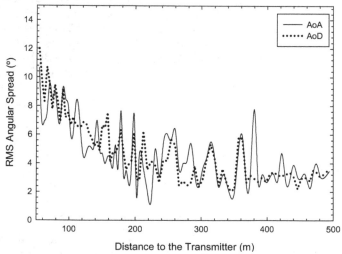

Fig. 9. Angular spread of the DOA/DOD of the rays in a road tunnel (Garcia-Pardo et al., 2011).

Distance (m)	100	200	300	400	500
Mean delay spread (ns)	2.5	1.7	2	1.8	2

Table 3. Mean delay spread in a semi arched road tunnel.

The fact that the delay spread remains constant can thus be explained by this decrease of A_s leading to constant time intervals between successive rays. Other measurements carried out in curved tunnels or tunnels presenting a more complex structure are reported for example in (Wang & Yang, 2006; Siy et al., 2009).

4. MIMO communications in tunnels

To improve the performances of the link, one of the most effective approaches recently developed is based on MIMO techniques (Foschini & Gans, 1998). Enhancement of spectral efficiency and/or decrease of the bit error rate (BER) were clearly emphasized for indoor environments, but in this case the paths between the Tx and Rx array elements are not strongly correlated (Correia, 2006). In a tunnel, however, the DOA/DOD of the rays are not widely spread, and one can wonder whether the number of active modes produce both a sufficient spatial decorrelation and a distribution of the singular values of the H matrix so that space time coding will yield a significant increase in the channel capacity.

4.1 Prediction of capacity from propagation models

To compute the elements of the H matrix, ray theory can be applied by adding all complex amplitudes of the rays received at each Rx antenna, this process being repeated for all Tx antennas. However, the computational cost increases when the receiver is placed far from Tx since the number of reflections needed to reach convergence becomes very large. A more interesting approach is the modal theory (Kyritsi & Cox, 2002; Molina-Garcia-Pardo et al., 2008a). In an $M \times N$ MIMO system (M being the number of Tx antennas and N the number of Rx antennas), the $N \times 1$ received signal $\vec{y}$ is equal to:

$$\vec{y} = H\vec{x} + \vec{n} \qquad (7)$$

where $\vec{x}$ is the $M \times 1$ transmitted vector and $\vec{n}$ is the $N \times 1$ additive white Gaussian noise vector. The transfer matrix H is fixed, i.e. deterministic, for any given configuration. The capacity for a given channel realization is given by (Telatar, 1995; Foschini & Gans, 1998):

$$C = \log_2\left(\det\left(I_N + \frac{SNR}{M}HH^\dagger\right)\right) = \sum_{i=1}^{\min(N,M)} \log_2\left(1 + SNR\,\lambda_i\right) \qquad (8)$$

where I_N is the NxN identity matrix, $\dagger$ represents the conjugate transpose operation, λ_i are the normalized eigenvalues of $HH^\dagger$ and SNR is the signal-to-noise ratio at the receiver. Let $A_{mn}^j(z)$ be the amplitude of the mode produced by the jth transmitting array element at the receiving axial location z. Each term $h_{ij}(z)$ of H, is the transfer function between the Tx element j and the Rx element i. This transfer function can be easily determined from (3):

$$h_{ij}(z) = \sum_{j=1}^{M}\sum_{m}\sum_{n} A_{mn}^{j}(z)\, e_{mn}(x_i, y_i) = \sum_{j=1}^{M}\sum_{m}\sum_{n} A_{mn}^{j}(0)\, e_{mn}(x_i, y_i)\; e^{-\gamma_{mn}z} \qquad (9)$$

In the above summation, the two terms $A_{mn}^{j}(z)$ and $e_{mn}(x_i, y_i)$ are functions of the position of the Tx element and of the Rx element of coordinates (x_i, y_i), respectively. We note that if the two sets of modes excited by two transmitting elements j_1 and j_2 have the same relative weight, two columns of H become proportional. In this case, H is degenerated and spatial multiplexing using these elements is no longer possible. In (Molina-Garcia-Pardo et al., 2008a), two rectangular tunnels were considered, with transverse dimensions 8 m × 4.5 m (large tunnel) and 4 m × 4.5 m (small tunnel). One conclusion of this work is that the Tx antenna must be off-centered, so that a large number of modes are excited with nearly similar weights, contrary to the case of an excitation by a centered element. Numerical applications show that, at a distance of 600 m and for an excitation by an antenna offset of 1/4 of the tunnel width, there are five modes whose relative amplitudes, when referred to the most energetic mode, are greater than -7 dB, whereas for the centered source, there are only three modes. In the following, we thus consider off-centered source positions, in order to excite a large number of active modes. Another limiting factor in MIMO performance is the correlation between receiving array elements (Almers et al., 2003). As an example, assume that the transmitting array is situated at 1/4 of the tunnel width and at 50 cm from the ceiling. For the two previous tunnels, we calculate the average spacing Δx that produces a correlation coefficient ρ, between vertical electric fields at x and $x+\Delta x$, smaller or equal to 0.7. Results in Fig. 10 show that the correlation distance is an increasing function of the axial distance, due to the decrease in the number of active modes.

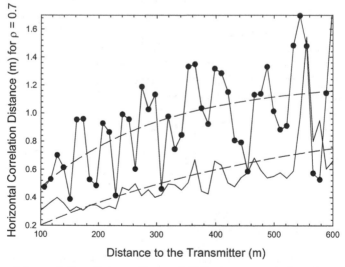

Fig. 10. Horizontal correlation distance (for $\rho = 0.7$) for two tunnels for a frequency of 900 MHz. Curve with successive points deals with the tunnel 8 m wide, the other curve corresponds to the tunnel 4.5 m wide, the dashed lines give the averaged values (Molina-Garcia-Pardo et al., 2008a).

One can also calculate the correlation $\rho^E_{j_1j_2}(z)$ between the electric fields received in a transverse plane, when the transmitting elements are j_1 and j_2, successively. Mathematical expressions of this correlation function are given in (Molina-Garcia-Pardo et al., 2008a), these authors also introduce a correlation function between modes. An example is given in Fig. 11, where the variation of $\rho^E_{j_1j_2}(z)$ for the large tunnel is plotted. We observe that, as expected, $\rho^E_{j_1j_2}(z)$ increases with distance. It is strongly dependent on the spacing between the Tx elements, and especially when this spacing becomes equal or smaller than 2λ.

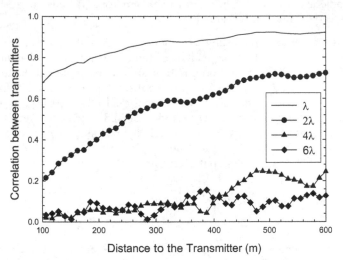

Fig. 11. Average correlation between the electric fields produced by two transmitters in the large tunnel for different array element spacing (6λ, 4λ, 2λ and λ) (Molina-Garcia-Pardo et al., 2008a).

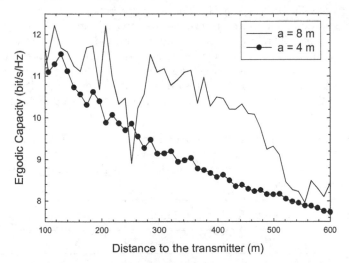

Fig. 12. Capacity of a 4x4 MIMO system for a SNR of 10 dB in both tunnels (Molina-Garcia-Pardo et al., 2008a).

Finally, the theoretical ergodic capacity C, calculated for an SNR of 10 dB versus distance, is presented in Fig. 12 for the two tunnel widths. At short distances, important fluctuations in C are found in the large tunnel due to the great number of active modes. In the small tunnel, due to the reduction in the number of modes and of correlation effects, C is smaller. For the same reason, we observe that in both tunnels, C is a continuously decreasing function of distance. We have thus seen that the concept of spatial diversity usually used in MIMO must be replaced, in tunnels, by the concept of modal diversity, the problem being to optimize the antenna array to excite numerous modes and thus to guarantee a low correlation between the multipath channels.

To conclude this section, it is interesting to have a measure of the multipath richness of the tunnel, and thus of the capacity, independent of the number of array elements. Therefore, we define a reference scenario corresponding to a uniform excitation of the tunnel and to a recovery of all modes in the Rx plane. This can be theoretically achieved by putting, in the whole transverse plane of the tunnel, a two-dimensional (2D) equal-space antenna array with inter-element spacing smaller that $\lambda/2$ (Loyka, 2005). In such a reference scenario, even if unrealistic from a practical point of view, each eigenvalue λ_i corresponds to the power P_{mn} of a mode EH_{mn} in the receiving plane (Molina-Garcia-Pardo et al., 2009a, 2009b). In this case, the capacity $C^{ref\,sce}$ in this scenario is given by:

$$ C^{ref\,sce} = \lim_{k \to \infty} \left[\sum_{i=1}^{k} \log_2\left(1 + SNR\,\Lambda_i^{ref\,sce}\right) \right] \quad \text{with} \quad \Lambda_i^{ref\,sce} = \frac{\exp\left(2\left(\alpha_{11} - \alpha_{mn}\right)z\right)}{\sum\limits_{m,n} \exp\left(2\left(\alpha_{11} - \alpha_{mn}\right)z\right)} \quad (10) $$

$C^{ref\,sce}$ only depends on the SNR and on the attenuation constants α_{mn} of the modes and thus on the frequency and on the electrical and geometrical characteristics of the tunnel.

Fig. 13 shows the variation in $C^{ref\,sce}$ versus the number k of eigenvalues of H or of the modes taken into account, the modes being sorted in decreasing power, for different frequencies between 450 and 3600 MHz. The distance between Tx and Rx is 500 m. At 450 MHz, only one mode mainly contributes to the capacity, and at 900 MHz the capacity still rapidly converges to its real value, since high order modes are strongly attenuated. At 1800 MHz, the summation must be made on at least 10 modes, since the attenuation constant in a smooth rectangular tunnel decreases with the square of frequency. At 3600 MHz, the attenuation of the first 30 modes is rather small and it appears that the curve nearly fits the curve "Rayleigh channel" plotted by introducing i.i.d. values as entries for the H matrix.

Finally, for this reference scenario and for a frequency of 900 MHz, Table 4 gives the number of modes needed to reach 90% of the asymptotic value of the capacity at different distances. It clearly shows the decreasing number of modes playing a leading part in the capacity when increasing the distance.

Distance Tx-Rx	d = 125 m	d = 250 m	d = 500 m	d = 1000 m
Number of modes	8	5	3	2

Table 4. Number of modes needed to reach 90% of the asymptotic value of the capacity at 900 MHz and for different distances.

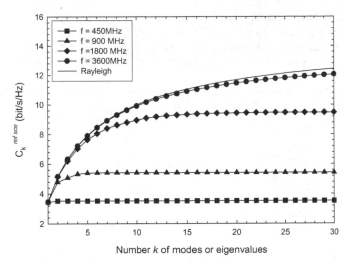

Fig. 13. Capacity $C_k^{ref\,sce}$ for the reference scenario versus the number k of eigenvalues or modes taken into account. The distance between Tx and Rx is 500 m, the tunnel being 5 m wide and 4 m high (Molina-Garcia-Pardo et al., 2009b).

To sum up this section, let us recall that the excitation of the modes plays a leading part in MIMO systems. It is thus interesting to excite the maximum number of modes, with nearly the same amplitude at the receiver plane. Optimum array configurations must thus be designed to obtain the maximum profit from MIMO systems. Furthermore, changes to the tunnel shape, such as narrowing or curves, will modify the weight of the modes and consequently the MIMO performance, as we will see in the next section.

4.2 MIMO channel capacity determined from experimental data

As stated before, many experiments have been conducted in mines and tunnels to extract both narrowband and wideband channel characteristics, but MIMO aspects were only recently considered. To illustrate both the interest of using MIMO in tunnels and the limitation of this technique, two scenarios are successively considered. First, the expected capacity will be determined from measurement campaigns in a subway tunnel whose geometry is rather complicated. Then we will consider the straight road arched tunnel discussed in 3.1.2, and whose photo is given in Fig. 4, to discuss the interest or not of using polarization diversity in conjunction with MIMO systems. Other studies, published in the literature, deal with pedestrian tunnels (Molina-Garcia-Pardo et al., 2003, 2004) where the position of the outside antenna is critical for the performance of the system. More recently, investigations into higher frequency bands have been carried out either for extracting double directional characteristics in road tunnels (Siy Ching et al., 2009), or for analyzing the performance of MIMO in the Barcelona Metro (Valdesuerio et al., 2010).

4.2.1 MIMO capacity in a subway tunnel

Measurements were made along the subway line, shown in Fig. 14, between two stations, Quai Lilas and Quai Haxo, 600 m apart (Lienard et al., 2003). The geometrical configuration can be divided into two parts: Firstly, there is a two-track tunnel that exhibits a significant curve along 200 m, from Lilas (point A) to point B, and then the tunnel is straight from point B to point D (100 m apart). Beyond this point and up to Haxo, the tunnel is narrow, becoming a one-track tunnel for the last 300 m (from D to C). The width of the one-track and of the two-track tunnel is 4 m and 8 m, respectively. The height of the tunnel is 4.5 m. Its cross-section is arched, but numerous cables and equipment are supported on the walls.

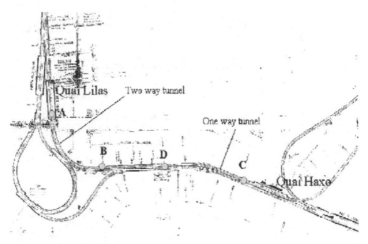

Fig. 14. Plan view of the tunnel. The distance between Quai Lilas and Quai Haxo is 600 m.

For studying a 4x4 MIMO system, 4 horn antennas were located on the platform. On the train, due to operational constraints, the patch antennas had to be placed behind the windscreen. In order to minimize the correlation, these antennas were placed at each corner of the windscreen. The channel sounder was based on a correlation technique and at a center frequency of 900 MHz and a bandwidth of 35 MHz. Both theoretical and experimental approaches have shown that, within this bandwidth, the channel is flat. The complex channel impulse response can thus only be characterized by its complex peak value and the MIMO channel is described in terms of H complex transfer matrices. All details on these measurements can be found in (Lienard et al., 2003).

The orientation of the fixed linear array inside a tunnel is quite critical. Indeed, the correlation distance between array elements is minimized when the alignment of the array elements is perpendicular to the tunnel axis. This result can be easily explained from the interference between active modes giving a signal which fluctuates much more rapidly in the transverse plane than along the tunnel axis. From the measurement of the H matrices, the expected capacity C was calculated for any position of the train moving along the tunnel. If we consider the first 300 m, the fixed array being placed at Quai Lilas, the propagation occurs in the two-track tunnel. Curves in Fig. 15 represent the variation in the cumulative distribution function of C using either a single antenna in transmission and in reception (SISO) or a simple diversity in reception (Single Input Multiple Output – SIMO) based on the maximum ratio combining technique or a 4x4 MIMO technique. These capacities can also be compared to those which would be obtained in a pure Rayleigh channel. In all cases, the signal to noise ratio (SNR) is constant, equal to 10 dB. An increase in capacity when using MIMO is clearly shown in Fig. 15.

Indeed, for a probability of 0.5, C is equal to 2, 5 and 9 bit/s/Hz, for SISO, SIMO and MIMO, respectively. In a Rayleigh environment, C would be equal to 11 bit/s/Hz.

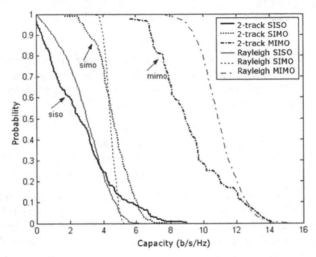

Fig. 15. Cumulative distribution of the capacity in a two-way tunnel for different diversity schemes, for an *SNR* of 10 dB (Lienard et al., 2003).

When the mobile enters the one-track tunnel at point D, one can expect an important decrease in the capacity, even by keeping the same *SNR*. Indeed, as we have seen in 2.1, the high order modes will be strongly attenuated if the tunnel narrows, leading to a decrease in the number of active modes and thus to an increase in the correlation between elements of the mobile array. In Fig. 16, we see that the capacity, in the order of 9 bit/s/Hz in the two-

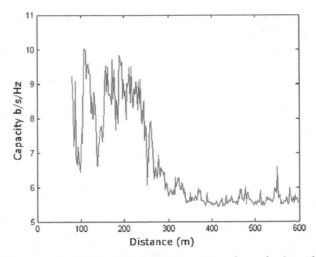

Fig. 16. Capacity for a constant SNR of 10 dB. Transmission from the 2-track tunnel, reception first in the 2-track tunnel (0 - 300 m), and then in the 1-track tunnel (300m – 600m). (Lienard et al., 2006).

track tunnel, decreases to 5.5 bit/s/Hz after the narrowing and thus reaches the capacity of a SIMO configuration. A more detailed explanation based on the correlation coefficients is given in (Lienard et al., 2006).

4.2.2 MIMO capacity in a straight road tunnel and influence of polarization diversity

Let us now consider the straight arched tunnel presented in 3.1.2 and a 4x4 MIMO configuration. In the experiments briefly described in this section, but detailed in (Molina-Garcia-Pardo et al., 2009c), the total length of each array is 18 cm. Different orientations of the elements, i.e. different polarizations, were considered:

1. VV: All Tx and Rx elements are vertically polarized.
2. HH: All Tx and Rx elements are horizontally polarized.
3. VHVH: Both the first and the third elements of each array are vertically polarized, while the second and the fourth elements are horizontally polarized.

In these 3 cases the inter-element spacing is thus equal to 6 cm. Another configuration, called "Dual", because Tx and Rx elements are dual-polarized, was also studied. In this case, the length of the array can be reduced from 18 cm to 6 cm. Channel matrix measurements were made in a frequency band extending from 2.8 to 5 GHz. Since the propagation characteristics do not vary appreciably in this band, average values of the capacity were determined from the experimental data, always assuming a narrow band transmission, i.e. a flat channel. The mean capacity C is plotted in Fig. 17 assuming a constant SNR of 15 dB, whatever the location of the mobile array. The worst configuration is VV or HH, and is due to the decrease in the number of active modes at large distances, leading to an increase in the correlation between array elements, as previously explained. For VHVH, C does not appreciably vary with distance and remains in the order of 16 bit/s/Hz. This better result comes from the low correlation between cross-polarized field components, the distance

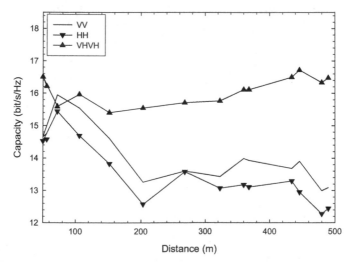

Fig. 17. Mean MIMO capacity assuming a fixed SNR at the receiver of 15 dB (Molina-Garcia-Pardo et al., 2009c).

between co-polarized array elements also increasing from 6 cm to 12 cm. A similar result was obtained with the Dual configuration. As a comparison, the theoretical capacity of a 4x4 MIMO in a i.i.d. Rayleigh channel would be around 20 bit/s/Hz, while for a SISO link C would be 5 bit/s/Hz, thus much smaller than for MIMO.

These results give some insight into the influence of the number of active modes and of the correlation between array elements on the capacity inside a tunnel. However, in practice, the Tx power is constant and not the SNR. Let us thus assume in a next step a fixed Tx power, its value being chosen such that at 500 m and for the VV configuration, an SNR of 15 dB is obtained. The capacities for the different array configurations and given in Fig. 18 do not differ very much from one another. This result can be explained by taking both the X-polar discrimination factor and attenuation of the modes into account. First we observe that VV gives slightly better results than HH, the attenuation of the modes corresponding to the vertical polarization being less significant. If we now compare HH and VHVH, we find that the two curves are superimposed. Indeed, we are faced with two phenomena. When using VHVH the spacing between co-polarized array elements is larger than for HH (or VV) and, consequently, the correlation between array elements is smaller. Unfortunately, the waves remain polarized even at large distances, as outlined in Table 3 of 3.1.2. This means that the signal received on a vertical Rx element comes only from the two vertical Tx elements and not from the horizontal Tx elements.

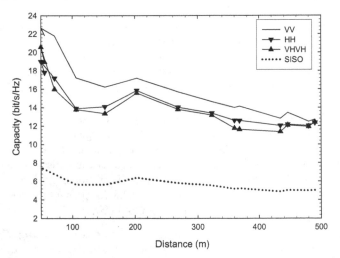

Fig. 18. Mean MIMO capacity assuming a fixed Tx Power (Molina-Garcia-Pardo et al., 2009c).

The total received power in the VHVH configuration is thus nearly half the total power in the HH (or VV) case. It seems that, comparing VHVH and HH, the decrease in power is compensated for by the decrease in correlation, leading to the same capacity. A similar result has been obtained when calculating the capacity of the Dual array. In all cases, the capacity obtained with such MIMO configurations is much larger that the SISO capacity for vertical polarization, as also shown in this figure. In conclusion, changing the polarization of the successive elements of the antenna arrays, as for VHVH, does not result in an improvement in MIMO performance in an arched tunnel. The advantage of the Dual configuration is that it decreases the length of the array while keeping the same channel capacity.

4.3 Performances of MIMO communications schemes

In previous sections, propagation aspects and capacity have been studied, but the bit error rate (BER) is one of the most important system design criterion. The robustness of MIMO to cope with the high correlation between array elements strongly depends on the MIMO architecture. MIMO can be used in three ways: beamforming, spatial multiplexing and space-time coding. Beamforming is useful for increasing the SNR and reducing the interference, spatial multiplexing increases the throughput by transmitting the independent flow of data on each antenna, while space time codes decrease the BER. In the following, we will consider space time coding and spatial multiplexing and we will compare the robustness of two well-known architectures, namely Vertical Bell Laboratories Layered Space-Time – VBLAST - (Wolniansky et al., 1998) and Quasi-Orthogonal Space-Time Block Codes – QSTBC - (Tirkkonen et al., 2000; Mecklenbrauker et al., 2004; Tarokh, 1999), assuming a narrow band transmission, i.e. a flat channel. The symbol detection method is based on the Minimum Mean Squared Error (MMSE) algorithm.

The BER was determined from the measurement of the H matrices in the arched tunnel presented in 3.1.2. To be able to carry out a statistical approach, the BER was calculated for numerous Rx locations in the tunnel and for 51 frequencies equally spaced in a 70 MHz band around 3 GHz. In (Sanchis-Borras et al., 2010), two transmission zones were considered: One near Tx, between 50 m and 150 m, and one far from Tx, between 400 and 500 m. In the following, only the results far from Tx are presented. The statistics on the BER were calculated owing to a simulation tool of the MIMO link and by considering 100 000 transmitted symbols, leading to a minimum detectable BER of 10^{-5}. The Tx power is assumed to be constant and was chosen such that a SNR of 10 dB is obtained at 500 m for the VV configuration. To make a fair comparison between MIMO and SISO, chosen as a reference scenario, the throughput and the transmitting power are kept constant in all cases. This means that the modulation schemes are chosen in such a way that the bit rate is the same for SISO and MIMO. In our examples, we have thus chosen a 16QAM scheme for SISO and QSTBC, and a BPSK for 4x4 VBLAST. The complementary cumulative distribution functions (ccdf) of the BER were calculated for the various transmission schemes, MIMO-VBLAST, MIMO-QSTBC and SISO, for two array configurations, VV and Dual. We have chosen these two kinds of array since, as shown in 4.2.2, they present the best performances of capacity under the assumption of constant transmitted power.

The results, presented in Fig. 19, show that there is no benefits from using VBLAST for the VV antenna configuration since the BER (curve 1) does not differ from the BER of a SISO link (curve 5). Despite the fact that the received power with co-polarized Tx and Rx arrays is maximized, the important correlation between the nearest antennas of the Tx/Rx arrays gives rise to a strong increase in the BER. Such a sensitivity of VBLAST to the correlation between antennas was already outlined in (Xin & Zaiping, 2004).

If VBLAST is used, but with dual-polarized antennas (Fig. 19, curve 2), the decrease in correlation allows for a better performance despite the fact that the average received power is smaller. We also note that QSTBC always gives better results than VBLAST. For this QSTBC scheme, contrary to what occurs for VBLAST, the BER is slightly better for VV polarization, i.e. for a higher received power, with QSTBC being much less sensitive to the correlation between antennas than VBLAST.

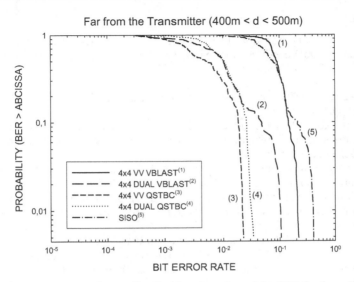

Fig. 19. Complementary cumulative distribution function of the BER far from the transmitter and for a fixed transmitting power (Sanchis-Borras et al., 2010).

As a comparison, we have also calculated the BER for a Rayleigh environment, assuming the same average SNR as for VV in tunnel. Dual is not compared to Rayleigh because in tunnel the waves are strongly polarized leading to a non-uniform distribution of the H matrix elements, in contrast to the case of a pure Rayleigh distribution. It appears that results obtained with QSTBC in tunnels are close to those which would be obtained in a Rayleigh environment, even at great distances from Tx. In conclusion, for a communication link in tunnels, MIMO outperforms SISO not only in terms of mutual information but also in terms of BER, assuming of course the same transmitting power and the same throughput, and under the condition that the number of active modes in the receiving plane is sufficient. Due to the guided effect of the tunnel and the attenuation of high order propagating modes at a large distance from the transmitter, the correlation between array elements strongly increases with distance and the QSTBC transmission scheme is thus more appropriate.

5. Concluding remarks

From measurements carried out in a straight tunnel of arched cross-section, which is quite a usual shape, we have shown that experimental results can be interpreted by means of an equivalent rectangular tunnel. The relevance of using MIMO techniques in tunnels was then investigated by first studying the correlation between array elements and the properties of the MIMO channel transfer matrix. This has been done by introducing the concept of active modes existing in the receiving plane. Despite the fact that the waves are guided by the tunnel, leading to a small angular spread for the paths relating the transmitter and the receiver, it appears that MIMO improves the channel capacity owing to a so-called modal diversity. Results obtained in more complex environments were also presented, such as in subway tunnels. In terms of bit error rate with space-time coding, a transmission scheme robust against antenna correlation must be chosen.

6. References

Almers, P.; Tufvesson; F., Karlsson, P. & Molisch, A. (2003). The effect of horizontal array orientation on MIMO channel capacity, *Proc. of the IEEE Vehicular Techno. Conf.*, Vol. 1, pp. 34–38, Seoul, Corea, April 22-25

Arshad, K.; Katsriku, F. A. & Lasebae, A. (2008). Modelling obstructions in straight and curved rectangular tunnels by finite element approach, *J. of Electrical Eng.*, Vol. 59, No. 1, pp. 9-13

Boutin, M.; Benzakour, A.; Despins, C. L. & Affes, S. (2008). Radio wave characterization and modeling in underground mine tunnels, *IEEE Trans. on Antennas and Propag.*, Vol. 56, No. 2, pp. 540 – 549

Chen, S. H. & Jeng, S. K. (1996). BR image approach for radio wave propagation in tunnels with and without traffic, *IEEE Trans. on Vehicular Techno.*, Vol. 45, No. 3, pp. 570-578

Chen, C. H.; Chiu, C. C.; Hung, S. C. & Lin, C.H. (2004). BER performance of wireless BPSK communication system in tunnels with and without traffic, *Wireless Personal Commun.*, Vol. 30, No. 1, pp. 1-12

Correia, L. M. (2006). *Mobile broadband multimedia networks*, Elsevier

Didascalou, D.; Maurer, J. & Wiesbeck, W. (2001). Subway tunnel guided electromagnetic wave propagation at mobile communications frequencies, *IEEE Trans. on Antennas and Propag.*, Vol. 49, No. 11, pp. 1590-1596

Dudley, D. G. (1994). *Mathematical Foundations for Electromagnetic Theory*, IEEE Press, New York

Dudley, D. G.; Lienard, M.; Mahmoud, S. F. & Degauque, P. (2007). Wireless propagation in tunnels, *IEEE Antennas and Propag. Mag.*, Vol. 49, No. 2, pp. 11–26

Emslie, A. G.; Lagace, R. L. & Strong, P. F. (1975). Theory of the propagation of UHF radio waves in coal mine tunnels, *IEEE Trans. on Antennas and Propag.*, Vol. 23, No. 2, pp. 192-205

Foschini, G. J. & Gans, M. J. (1998). On limits of wireless communications in a fading environment when using multiple antennas, *Wireless Personal Commun.*, Vol. 6, No. 3, pp. 311–335

Garcia-Pardo, C.; Molina-Garcia-Pardo, J. M.; Lienard, M. & Degauque, P. (2011). Time domain analysis of propagation channels in tunnels, *Proc. of the 7th Advanced Int. Conf. on Commun.*, St. Maarten, The Netherlands Antilles, March 20 – 25

Hrovat, A.; Kandus, G. & Javornik, T. (2010). Four-slope channel model for path loss prediction in tunnels at 400 MHz, *IET Microw. Antennas Propag.*, , Vol. 4, No. 5, pp. 571–582

Kyritsi, P. & Cox, D. (2002). Expression of MlMO capacity in terms of waveguide modes, *Electronic Letters*, Vol. 38, No. 18, pp. 1056-1057

Laakmann, K. D. & Steier, W. H. (1976). Waveguides: characteristic modes of hollow rectangular dielectric waveguides, *Appl. Opt.*, Vol. 15, pp. 1334–1340

Lienard, M.; Lefeuvre, P. & Degauque, P. (1997). Remarks on the computation of the propagation of high frequency waves in a tunnel, *Annals Telecomm.*, Vol. 52, No. 9-10, pp. 529-533

Lienard, M. & Degauque, P. (1998). Propagation in wide tunnels at 2 GHz: A statistical analysis, *IEEE Trans. on Vehicular Techno.* Vol. 47, No. 4, pp. 1322 - 1328

Lienard, M. & Degauque, P. (2000a). Natural wave propagation in mine environments, *IEEE Trans. on Antennas and Propag.*, Vol. 48, No 9, pp 1326 – 1339

Lienard, M. ; Betrencourt, S. & Degauque, P. (2000b). Propagation in road tunnels: a statistical analysis of the field distribution and impact of the traffic, *Annals of Telecom.*, Vol. 55, No. 11-12, pp. 623-631

Lienard, M.; Degauque, P.; Baudet, J. & Degardin, D. (2003). Investigation on MIMO channels in subway tunnels, *IEEE J. on Selected Areas in Commun.*, Vol. 21, No. 3, pp. 332–339

Lienard, M.; Degauque, P. & Laly, P. (2004). A novel communication and radar system for underground railway applications, *European J. of Transport and Infrastructure Research*, Vol. 4, No. 4, pp. 405-415

Lienard, M.; Degauque, P. & Molina-Garcia-Pardo, J. M. (2006). Wave propagation in tunnels in a MIMO context - A theoretical and experimental study, *Comptes Rendus. Phys.*, Vol. 7, pp. 726-734

Loyka, S. (2005). Multiantenna capacities of waveguide and cavity channels, *IEEE Trans. on Vehicular. Techno.*, Vol. 54, pp. 863–872

Mahmoud, S. F. & Wait, J. R. (1974). Geometrical optical approach for electromagnetic wave propagation in rectangular mine tunnels, *Radio Science*, Vol. 9, pp. 1147–1158

Mahmoud, S. F. (1988). *Electromagnetic waveguides theory and applications*, Peter Pergrenus Ed.

Mahmoud, S. F. (2008). Guided mode propagation in tunnels with non-circular cross section, *Proc. of the IEEE Int. Symp. on Antennas and Propag.*, San Diego, USA, July 5-11

Mahmoud, S. F. (2010). *Wireless transmission in tunnels*, Chapter 1 in the book *Mobile and Wireless Communications*, InTech, ISBN 978-953-307-043-8

Mariage, P.; Lienard, M. & Degauque, P. (1994). Theoretical and experimental approach of the propagation of high frequency waves in road tunnels, *IEEE Trans. on Antennas and Propag.*, Vol. 42, No. 1, pp. 75-81

Marti Pallares, F.; Ponce Juan, F. J. & Juan-Llacer, L. (2001). Analysis of path loss and delay spread at 900 MHz and 2.1 GHz while entering tunnels, *IEEE Trans. on Vehicular Techno.*, Vol. 50, No.3, pp. 767-776

Mecklenbrauker, C. F. & Rupp, M. (2004). Generalized Alamouti codes for trading quality of service against data rate in MIMO UMTS, *EURASIP J. on Applied Signal Processing*, No. 5, pp. 662–675

Molina-Garcia-Pardo, J. M.; Rodríguez, J. V. & Juan-Llácer, L. (2003). Angular Spread at 2.1 GHz while entering tunnels, *Microwave and Optical Technology Letters*, Vol. 37, No. 3, pp. 196-198

Molina-García-Pardo, J. M.; Rodríguez, J. V. & Juan-Llacer, L. (2004). Wide-band measurements and characterization at 2.1 GHz while entering in a small tunnel, *IEEE Trans. on Vehicular Techno.*, Vol. 53, No. 6, pp. 1794-1799

Molina-Garcia-Pardo, J. M.; Lienard, M.; Degauque, P.; Dudley, D. G. & Juan Llacer, L. (2008a). Interpretation of MIMO channel characteristics in rectangular tunnels from modal theory, *IEEE Trans. on Vehicular Techno.*, Vol. 57, No. 3, pp. 1974-1979

Molina-Garcia-Pardo, J. M.; Nasr, A.; Liénard, L.; Degauque, P. & Juan-Llácer, L. (2008b). Modelling and understanding MIMO propagation in tunnels, *Proc. of 2nd International Conf. on Wireless Communications in Underground and Confined Areas* Val-d'Or, Québec, Canada, August 25-27

Molina-Garcia-Pardo, J. M.; Lienard, M.; Nasr, A. & Degauque, P. (2008c). On the possibility of interpreting field variations and polarization in arched tunnels using a model for propagation in rectangular or circular tunnels, *IEEE Trans. on Antennas and Propag.*, Vol. 56, No. 4, pp. 1206–1211

Molina-Garcia-Pardo, J. M.; Lienard, M.; Stefanut, P. & Degauque, P. (2009a). Modelling and understanding MIMO propagation in tunnels, *J. of Communications*, Vol. 4, No. 4, pp. 241-247

Molina-Garcia-Pardo, J. M.; Lienard, M.; Degauque, P.; Simon, E. & Juan Llacer, L. (2009b) On MIMO channel capacity in tunnels, *IEEE Trans. on Antennas and Propag.*, Vol. 57, No. 11, pp. 3697-3701

Molina-Garcia-Pardo, J. M.; Lienard, M.; Degauque, P.; García-Pardo, C. & Juan-Llacer, L. (2009c). MIMO channel capacity with polarization diversity in arched tunnels, *IEEE Antennas and Wireless Propagation Letters*, Vol. 8, pp. 1186-1189

Molina-Garcia-Pardo, J. M.; Lienard, M. & Degauque, P. (2009d). Propagation in tunnels: experimental investigations and channel modeling in a wide frequency band for MIMO applications, *EURASIP J. on Wireless Commun. and Networking*, Article ID 560571, 9 pages, doi:10.1155/2009/560571

Molisch, A. F. (2005), *Wireless communications*, IEEE Press

Popov, A. V. & Zhu, N. Y. (2000). Modeling radio wave propagation in tunnels with a vectorial parabolic equation, *IEEE Trans on Antennas and Propag.*, Vol. 48, No. 9, pp. 1316 - 1325

Rappaport, T. S. (1996). *Wireless communications: Principles and practice*, Prentice Hall Ed.

Sanchis-Borras, C.; Molina-Garcia-Pardo, J. M.; Lienard, M.; Degauque, P. & Juan LLacer, L. (2010). Performance of QSTBC and VBLAST algorithms for MIMO channels in tunnels, *IEEE Antennas and Wireless Propag. Letters*, Vol. 9, pp. 906-909

Siy Ching, G.; Ghoraishi, M.; Landmann, M.; Lertsirisopon, N.; Takada, J.; Imai, T.; Sameda, L. & Sakamoto, H. (2009). Wideband polarimetric directional propagation channel analysis inside an arched tunnel, *IEEE Trans. on Antennas and Propag.*, Vol. 57, No. 3, pp. 760 - 767

Tarokh, V.; Jafarkhani, H. & Calderbank, A. R. (1999). Space-Time block codes from orthogonal designs, *IEEE Trans. on Information Theory*, Vol. 45, pp. 1456-1467

Telatar, I. E. (1995). Capacity of multi-antenna gaussian channel, *European Trans. on Telecom.*, Vol. 10, pp. 585–595

Therrien, C. W. (1992). *Discrete Random Signals and Statistical Signal Processing*, Prentice Hall, Englewood Cliffs, New Jersey

Tirkkonen, O.; Boariu, O. & Hottinen, A. (2000). Minimal nonorthogonality rate one space time block codes for 3+ Tx antennas, *Proc. of the IEEE 6th Int. Symp. on Spread-Spectrum Tech. & Appli.*, pp. 429 - 432, New Jersey, USA, Sept. 6-8

Valdesuerio, A.; Izquierdo, B. & Romeu, J. (2010). On 2x2 MIMO observable capacity in subway tunnels at X-Band: an experimental approach, *IEEE Antennas and Wireless Propag. Letters*, Vol. 9, pp. 1099-1102

Wait, J. R. (1962). *Electromagnetic waves in stratified media*, Pergamon Press

Wang, T. S. & Yang, C. F. (2006). Simulations and measurements of wave propagation in curved road tunnels for signals from GSM base stations, *IEEE Trans. on Antennas and Propag.*, Vol. 54, No. 9, pp. 2577-2584

Wolniansky, P. W.; Foschini, G. J.; Golden, G. D. & Valenzuela, R. A. (1998). V-BLAST: An architecture for realizing very high data rates over the rich-scattering wireless channel, *Proc. of the URSI Symposium on Signals Systems and Electronics*, pp. 295-300

Xin, L. & Zaiping N. (2004). Performance losses in V-BLAST due to correlation. *IEEE Antennas and Wireless Propagation Letters*, Vol. 3, pp. 291-294

Yamaguchi, Y.; Abe, T. & Sekiguchi, T. (1989). Radio wave propagation loss in the VHF to microwave region due to vehicles in tunnels, *IEEE Trans. on Electromagn. Compat.*, Vol. 31, No. 1, pp. 87-91

Zhang, Y. P.; Zheng, G. X. & Sheng, J. H. (2001). Radio propagation at 900 MHz in underground coal mines, *IEEE Trans. on Antennas and Propag.*, Vol. 49, No. 5, pp. 757–762

2

Digital Communication and Performance in Nonprofit Settings: A Stakeholders' Approach

Rita S. Mano

Department of Human Services, University of Haifa, Haifa
Israel

1. Introduction

The importance of digital communication (hereafter DC) is often related to higher performance in for-profits but less often in non-profit organizations. Organizations considered as innovative engage in new ideas, attitudes, or behaviors that depart from their previous routines or strategies. Such ideas are often transmitted by, or easily traced in DC and the information-technology media, but these advantages seem to have been neglected until recently in the nonprofit sector and outcomes following DC exposure are generally ambiguous (Sidel & Cour, 2003). This is probably because performance in nonprofits is not always clearly defined as the multiple stakeholders involved in nonprofit management have conflicting ideas about criteria of performance. Some suggest that social goals cannot be properly attained through DC whereas others show promising increases in economic goals - donations and participation- via digital sites. Not surprisingly, the disagreement as to whether DC definitely improves digital performance (hereafter DP) is an open field for investigation.

The present chapter elaborates on this aspect and attempts to develop a conceptual framework for assessing the contingent effects that mediate the DC - DP link (hereafter DCP) in nonprofit settings. We present a taxonomy based on the stakeholders' approach, followed by a conceptual model addressing: (a) type of organization; (b) type of stakeholder; and (c) type of performance. Lastly, we offer a set of propositions leading to the development of hypotheses for examining DCP links that enable formulation of empirical tests (Beckey, Elliot, & Procket, 1996).

DC enables information to be created, edited, stored, discarded or organized, and is easily accessed from relevant recipients or links in and between different facets of an organization, and/or social systems composed from multiple stakeholders. In nonprofits DC enables introducing and maintaining innovative services that increase social visibility (Corder, 2001; Balser & McKlusky, 2005; Herman & Renz, 2004). DC is therefore associated naturally with enhanced performance in nonprofits because "...by making that information available centrally, data can be consolidated, trends can be identified, and campaigns and resources can be more effectively focused..." (Burt & Taylor 2003, p. 125). Critics of the DC-DP link point though, that defining performance in nonprofits is problematic because the nonprofits rely - among other criteria – on the successful recruitment of social support which can only be attained through face to face interactions (Baruch & Ramalho, 2006; Mano, 2009, 2010).

Indeed the distinction between social and economic goals and their respective terms of effectiveness vs. efficiency suggest that the higher the number of supportive agents the higher that organizational effectiveness whereas organizational efficiency is linked to increasing economic returns to organizational objectives using internal and external resources to produce maximum results with a minimal investment.

As competition increases nonprofit organizations are often "trapped" by the importance of economic criteria of performance, pursuing efficiency (Galaskiewicz & Bielefeld, 1998). Efficiency though does not attest to the nature of nonprofit goals because the quality and importance of social goals is not easily measured and organizations can be efficient in certain dimensions and less efficient in others. As a result efficient operations reflect a limited and often distorted view of performance in nonprofits (Bielefeld (1992).

As remote communication services and the number of internet users increase nonprofits have turned to DC to improve their efficiency levels –lower costs- and effectiveness to increase social visibility by addressing potential audiences and improving relationships existing supporters –volunteers and funding- and develop new alliances. Some have found new efficiencies in their operations when advocating issues, as well as increased funding from virtually unchanged DC. In 2001 Olsen and Keevers made an exploratory study of how voluntary organizations employed 'button' options to enable different forms of contributing, e.g. Donate now; Donate; Donate Online; Giving, How you can help, Join or Give, Make a donation, Show your support. Similarly, the ePhilanthropy Foundation has developed the 'ePhilanthropy Toolbox' as a way of drawing attention to the wide array of techniques and services available. The elements of the toolbox fall into six categories, namely: Communication/Education and stewardship; Online donations and membership; Event registrations and management; Prospect research; Volunteer recruitment and management; and Relationship-building and advocacy (Hart, 2002).

With this taxonomy it is logical to assume that there is need to distinguish between different criteria of performance. In the present study we first distinguish between (a) effectiveness – attainment of organizational goals, and (b) efficiency – the economically "best" outcome for organizational operations; then we distinguish between two types of stakeholders: (a) public – recipients of services and providers of services at the national level; (b) private – individuals and organizational agents of support. Finally, we distinguish between individual level and organizational level outcomes of DC, applying these dimensions to two basic types of organization, i.e. service and advocacy groups.

A proposed model suggests that DC can give rise to different forms of DP (Seshadri, & Carstenson, 2007), but this necessitates adopting the suggestion that NPO performance is influenced by the different expectations of stakeholders. Accordingly, DC can either enhance effectiveness, i.e. attainment of the organization's social goals; or efficiency, i.e. attaining economically satisfactory outcomes.

2. Theoretical background

Internet has made a vast range of activities accessible to a huge number of individuals. In contrast to other types of social exchange, the internet is quick, efficient, and direct, so that minimal efforts are required to accomplish goals. DC is thus important for conveying information and increasing contact with stakeholders. Even though computer-mediated contact lacks the social and physical clues that are essential for good communication, individuals prefer to use DC as a medium with a social presence commensurate with the task they wish to accomplish (Hollingshead & Noshir, 2002), whether for employment, leisure, to broaden personal horizons, or for attaining social status (Cnaan, Handy & Wadsworth, 1996). These assumptions are at the core of the "functional" approach to DC that covers a wide range of aspects such as selective processing, channel complementation etc. (Dutta & Bodie, 2008). The common factor underlying this functional approach is "gratification", i.e. tailoring media features to individual needs.

Individuals who contribute to nonprofits may choose to make donations via DC, or participate in discussions and increase their levels of civil engagement. At the organizational level, it is clear that DC enables easy and quick access to information. In today's global economy, this is identified with improved organizational performance (Ducheneaut & Watts, 2005). Earlier studies in voluntary organizations in the United Kingdom (Burt & Taylor, 2003) showed increased use of DC to reconfigure key information flows in support of enhanced campaigning and more effective user services. There is also evidence that nonprofits are able to exploit opportunities for radical shifts in organizational arrangements and thereby increase their social, economic, and political spheres of influence. The authors pointed out that the use of advanced technological communication devices has improved both efficiency and potential by comparison with previous periods, as have levels of integration of information within organizations, while the level and quality of coordination have risen substantially as well. Combining individual levels of contribution through DC with organizational criteria of performance – DP - in nonprofit settings is thus possible as long as DC approaches the stakeholders' needs and expectations properly, so that nonprofit DP is possible by increasing either effectiveness or efficiency (Mano, 2009).

Stakeholders are those individuals or groups who support an organization. According to Abzug and Webb (1999), stakeholders of NPOs include the community, competitors, clients, managers and employees, government suppliers, and private founders of organizational goals. Researchers of NPOs (Freeman, 1984; Winn & Keller, 2001) stress the dependence of the organization on its stakeholders, namely any group or individuals that may influence or be influenced by an organization working to achieve its goals. Similarly, Donaldson and Preston (1995) claimed that stakeholders are the investors (of financial or human resources) who have something to lose or gain as a result of organizational activities. Others view stakeholders as those with power, having legitimate and justified claims on an organization.

The stakeholders approach involves three dimensions: first, type of performance – effectiveness vs..efficiency; second, two types of stakeholders, i.e. public – recipients and providers of services at the national level; and private – i.e. all individual and organizational agents of support that does not involve public funding; and third, individual and organizational outcomes of DC. We apply the above dimensions to two basic types of nonprofits, namely service and advocacy.

2.1 Digital performance - DP

To successfully achieve social visibility and contributions, nonprofits have been gradually and consistently entering the field of DC because it is easier and faster to: a) maintain contact with more stakeholders; (b) obtain access to immediately measurable outcomes.

Furthermore, it provides a sense of involvement to a large number of individuals and organizations (Bryson et al., 2001). A larger pool of supporters is more probable to promote secure and reliable resources despite economic fluctuations (Galaskiewicz & Bielefeld, 1998). However, among the many criteria of nonprofit performance raised over the years (Baruch & Ramalho, 2004), the focus is on two terms that define how nonprofits organize their activities i.e. *effectiveness* and *efficiency* (Herman & Renz, 2008).

2.2 Effectiveness vs. efficiency

Nonprofit organizations are social entities intended for defined tasks, and the concept of a "goal" is integral to their definition (Daft, 2004). Organizational success is measured in two main dimensions: effectiveness and efficiency, both exemplifying the degree to which an organization achieves its goals.

So what are those goals, and who defines them? Defining organizational goals in nonprofits is not simple because of the multiplicity of stakeholders. Some nonprofits seek goals related to the welfare of their clients (human service organizations), while others speak in the name of groups, communities and society in general (advocacy organizations). In both cases, effectiveness is evidence of the organization's ability to use its resources to achieve its unique goals. Human services attend to "individual" level needs, while advocacy organizations' goals are intended to generate support and emphasize importance to society (Perrow, 1961). Organizational effectiveness is therefore a "relative", even "ideology-driven" or "ethical" approach to the attainment of goals, to the extent that they are often debated and sometimes abandoned. By contrast, efficiency is a "technical" measure of performance focusing on nonprofits' dependence on resources and the importance of managing those resources. According to this concept, organizations must ultimately consider performance as based on quantitative measures – input versus output. So NPOs - like for-profit organizations - implement organizational tactics striving towards increased rationality and technical efficiency (Bozzo, 2000; Bryson, Gibbons, & Shaye, 2001; Carson, 2000). If, for each unit of output, there is a decrease in requested inputs, efficiency increases. DC, characterized by immediacy, speed and accuracy, can thus improve efficiency – level of donations, immediate enlistment of volunteers - lower investment – fewer people involved, fewer hours of work, etc. What can be confusing is not the level of performance, but how stakeholders evaluate each aspect of it. If, for example, nonprofits lose prestige but gain in volunteer enlistment by using DC, then they may become less attractive and lose social visibility, for example with groups concerned with deviant behavior. Situations in such cases are often insoluble, sometimes with conflicting views between stakeholders, in regard to evaluating whether effectiveness is more desirable than efficiency.

2.3 Private vs. public stakeholders

Mitroff (1983) emphasizes the difficulties arising from excessive numbers of stakeholders. The more stakeholders there are, the more complex are their demands, expectations and conditions (Freeman & Evan, 2001). On the basis of the analysis, activities are adjusted to suit the stakeholder groups. Lack of agreement between stakeholders becomes apparent when performance criteria must be established or negotiated (Alexander, 2000), for example, in questions such as who defines and who is responsible / accountable for delivery of a product or service. How NPO performance should be measured can arouse different opinions concerning those measurements (Rado et al., 1997; Mitchell, Aigle & Wood, 1997; Rowley, 1997). Some stakeholders adhere to the "technical" measures attached to cost-effectiveness, while others tend to emphasize ideological aspects and ignore economic measures of success, i.e. efficiency. However, recent economic measures of efficiency have become vital for the survival of NPOs (Shmidt, 2002; Bielefeld, 1992; Crittenden, 2000; Galaskiewicz & Bielefeld, 1998). Nonetheless, some authors suggest that it is difficult to define DP using a single measure, whether objective (Carson, 2000) or subjective (Baruch & Ramalho; 2004; Forbes, 1998; Herman & Renz, 2008).

2.4 Public stakeholders

Government support is an important resource, and for many nonprofits – mainly those providing human services - their dependence on grants and contracts from the government is their principal source of survival (Gronbjerg, 1993). This is due mainly to the recent uncertainty caused by the economic crisis, as well as to increased competition. According to Young (1999), there are three economic perspectives that can assist in understanding and analyzing the connection with the government, namely: the "supplementary" role of nonprofits for services that are not provided by the government; the "complementary" relationship of "contracts" in the provision of services; and the "adversarial" pressure on the government to change policies and laws (Salamon, 1995).

2.5 Private stakeholders

Private donations (from organizations and/or individuals) and grants from different foundations are traditional sources of support for third sector organizations (Froelich, 1999). However, as Gronbjerg (1992, 1993) pointed out, there is inherent fluctuation in the scope and steadiness of private donations. This lack of certainty impacts on the activities of an organization, and can even contribute to changes in organizational objectives. For instance, grants, which are characterized by professional teams and formal procedures, have strengthened the formalization of organizations in the third sector by fixing conditions for obtaining them, demanding accountings of implementation and evaluation of projects, goal definitions, etc. (Carson, 2000). Likewise, due to the need to comply with the wishes of contributors and thus ensure their support, third sector organizations change their organizational objectives in accordance with external pressures that are not necessarily related to organizational goals or to the receivers of services, thereby endangering their legitimacy (Galaskiewizc & Bielefeld, 1998). In order to maintain capital flow and increase income, third sector organizations must also provide incentives such as: promoting donors' prestige, commitment by the organization to serve the community, and participation of donors in achieving these goals and in making decisions related to the organization (Markham, Johson & Bonjean, 1999).

2.6 Individual vs. organizational stakeholders

DC has overall positive effects in many spheres and, despite alterations over time, individuals using internet today are more likely to evaluate DC as central to increased social involvement, civic participation, and community members' involvement. DC is often related to positive sentiments and a stronger sense of community "belonging". The contribution of DC was first examined in relation to media in local communities (Keith, Arthur, & Michael; 1997; Shah, Mcleod & Yoon, 2001), and was viewed as a means to increase social capital (Shah, Nojin & Lance, 2001). Recently, Mesch and Talmud (2010) re-assessed the impact of DC and found improvement in levels of satisfaction and commitment to the community. Using a longitudinal design of two suburban communities in Israel, they presented evidence of increased effectiveness of nonprofit settings – communities - using DC. Similar results were found by Postigo (2009), who examined volunteers demanding reasonable compensation in the USA, i.e. social factors and attitudes and a sense of community, creativity, and accomplishment were all related to participation in Internet forums, as was an increase in 'ethical consumption' among young people (Banaji & Buckingham, 2009). As a result, nonprofit marketing as well as commercial marketing through DC strives for clear recognition of differences in youth cultures to gradually enhance DP (Jennings & Zeitner, 2003). This is probably why, according to Tapia and Ortiz (2010), municipalities are attempting to create municipal-sponsored wireless broadband networks. The importance of increasing individual-level participation in local government is seen as a major goal in order to avoid wrong choices, failure, or mistrust of local governments by citizens.

To understand these effects on DCP we must also keep in mind that service organizations differ from advocacy groups because they address different types of stakeholders. Service organizations aim to achieve service in areas such as health, welfare and education, whereas the purpose of the advocacy groups is to be involved in national or international issues, presenting their ideas and participating in group discussions.

2.7 The DCP link

According to Hamburger (2008), there is high potential for nonprofit organizations using DC as a medium, to channel social development, because volunteers recruited through the net can contribute significantly to the lives of many millions of people in need throughout the world. Hamburger insists that the positive potential of the Internet derives from understanding the informative and communicative aspects of net-volunteering, in which three separate types need to be addressed, namely: the personal, i.e. to individual needs, the interpersonal, i.e. the positive effects of interaction; and the group or organizational level outcomes which address the goals.

An early study (Hart, 2002) suggest that the Internet is a reliable means for building support, providing a cost-effective opportunity to build and enhance relationships with supporters, volunteers, clients and the community they serve. Organizations have discovered that consistent and well targeted e-mail communication encourages users to access information that is central to gaining support. Referring to charities, the author recommends using it as both a communication and stewardship tool, and a fundraising tool, but warns that this is only an aid, and not a substitute for cultivating and enhancing relationships. Similar conclusions have been reached in regard to American service organizations (Olsen, Keevers, Paul & Covington, 2001).

Jennings & Zeitner's (2003) exploration of Internet use in regard to the engagement of American citizens between 1982 and 1997, demonstrated that the "digital divide" in civic engagement remained in place, indicating that Internet access has positive effects on several indicators of civic engagement, according to how much the Internet is used for political purposes. They concluded that Internet is effective only when personal political interest intersects with personal media familiarity. As the number of websites increases, (Hargittai, 2000) the battle to attract audiences may reduce the number of visitors who could possibly support nonprofits. This is apparently why recent studies focus on the DCP link. Barraket (2005) for example, in a content analysis of 50 Australian third- sector organizations' websites, showed use of DC and how such organizations are (or are not) utilizing online technology to attract citizen engagement, are compatible with their organizational activities. This is surprising but not irrational because, as suggested by Mano (2009), targeting techniques of nonprofits are lacking in sophistication. Recent evidence shows that the scope and intentions of nonprofits using the Internet are relatively blurred. Indeed, adopting a stakeholders' approach the author shows that using the Internet cannot be effective unless the communication style and targeting "fits" the needs of the targeted stakeholders. The study's empirical findings indicates that social marketing through DC increases support from private stakeholders, but decreases that of the public stakeholders.

Nevertheless, Bezmalinovic, Dhebar & Stokes (2008), evaluating the potential for successfully recruitment through online systems, show that technology is useful for selecting, recruiting and pooling volunteers, but that online processes also necessitate significant costs. Moreover, Finn, Maher and Forster (2006), using archival data, showed that DC has been the main vehicle for adoption and adoption-readiness for nonprofits in the United States, enabling them to take advantage of existing opportunities. Similarly, a growing field of studies in social marketing shows that technology can serve these organizational goals provided there is good fit between the medium used and its target (Mano, 2009). A recent study examining these aspects revealed that voluntary organizations now use much more technology to increase performance and sustain adaptation to environmental threats, but they still fall short of using the appropriate DC strategies to increase support for their cause (Mano, 2009). Similarly Suárez (2009), examining the social role of nonprofit e-advocacy and e-democracy (civic engagement) as identified by 200 nonprofit executive directors, revealed that rights groups, environmental organizations, and policy entrepreneurs are consistently likely to mention advocacy and promote civic engage-ment on their websites. Conversely, funding and/or dependence on resources generally fail to explain nonprofit use of websites for social purposes.

An organization can be effective in some dimensions but less efficient in Others and *vice versa*. Organizational efficiency is related to the accomplishment of organizational objectives and their cost, and to an organization's ability to use available resources to achieve maximal results with minimal investment. In commercial organizations, efficiency is usually measured in financial terms such as the annual level of sales per employee. In public service organizations, however, the budget is measured against the number of employees in the organization, or the number of clients dealt with per employee per week as a measure of output. The efficiency of an organization does not attest to the nature of its products, their quality, or the level of service it provides, and therefore only reflects a limited view of organizational success (Bielefeld, 1992). Unfortunately, nowadays nonprofit organizations become "trapped" by these "market" criteria, pursuing efficiency rather than effectiveness, which is the "wrong" way to achieve performance in nonprofit organizations (Galaskiewicz & Bielefeld, 1998). Reviewing these studies suggests that nonprofits are far more interested and adherent to effectiveness via DC, rather than efficiency. We combine the three proposed dimensions of the DCP link in Table 1.

	Type of Stakeholder	
Type of Organization	Private Stakeholders	Public Stakeholders
Service nonprofits	*Recipients*	*Government funded*
Advocacy nonprofits	*Participants*	*Social Institutions*

Table 1. A taxonomy of digital communication targets according to type of non profit organization and type of stakeholders

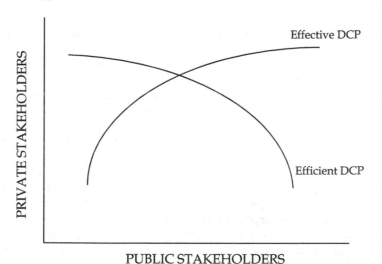

Fig. 1. Predicting Effective and Efficient DCP

Table 1 shows how nonprofits should treat stakeholders in order to obtain maximum effects of using DC. The model suggests that private stakeholders can be divided into two groups: a) recipients of services, and b) participants in the organization's online and offline activities. Public stakeholders are those who contribute and donate – money and in kind - for the benefit of the organization (Balser & McClusky. 2005). How DC improves performance in NPOs in fact represents how digitally mediated marketing affects expectations of stakeholders (Ackcin, 2001). Some recent studies show that, even though such differences are evident in the concept of "targeting", empirical evidence of how marketing in NP succeeds in targeting the "right" stakeholders (Vazquez, Ignacio, & Santos, 2002) and increases performance (Shoham, Ruvio, Vigoa-Gadot & Schwabsky, 2006) or whether levels of performance are adequately evaluated to reach conclusions, are scarce, because DC investments in marketing do not necessarily increase stakeholder support (Brainard & Siplon 2004; Vazquez, Ignacio & Santon, 2000; Mano, 2009;. 2010). How does this affect DCP?

Figure 1 presents a model to facilitate understanding of the DCP link. As suggested above, while efficient economic measures are vital (Schmidt, 2002; Bielefeld, 1992), the institutional nature of nonprofits suggests that effectiveness measures rise successfully when DC addresses social goals. Then general support is expected, such as increasing awareness of a cause rather than concrete outcomes such as donations. However, when DC is dealing with individual-level outcomes such as satisfaction, participation, or donations, then private stakeholders' expectations and needs should be taken into account. More attractive sites could be one way of gaining direct and/or indirect support (Crittenden, 2000; Galaskiewicz & Bielefeld, 1998; Mano, 2009).

3. Discussion

DC has made a great variety of activities accessible to a very large number of individuals and organizations. In contrast to other types of social exchange, the internet medium is quick, efficient, and direct, requiring minimum effort on the part of the user to accomplish goals, as opposed to traditional methods of interacting instead of "acts of reciprocity, negotiation and cooperation that once allowed businesses to build and maintain relationships gave way to optimized matching systems that undermined and diminished the role of trust" (Cheshire, Geabasi and Cook 2010, p.177). While it is clear that surmounting competition relies today on access and speed of access to sources of information, and the dissemination of that information to possible stakeholders, we must ask questions about the DCP link between digital communication and digital performance (Peltier & Schibrowski, 1995).

For nonprofits there is ambiguity about the extent to which digital communication enhances performance. There are some basic concerns about the extent to which nonprofits can successfully adapt tactics used by for-profit organizations for promotional goals since they often lack professional staff and there is a shortage of volunteers. In addition, nonprofits are examples of social entities that seek social goals, and standard digital communication often lacks the direct contacts considered as vital to social awareness. Moreover, nonprofit stakeholders do not agree about nonprofit performance. Some insist that nonprofits should not be concerned with the economic success of organizational activities, but with how much support they receive from social groups. It is less important whether they donate to a cause or even actively participate as members than being "aware of social goals" and embracing them. Nonetheless, board members and professional managers lacking continuous monetary support are confronted with difficulties in running the organization and completing projects, especially those which are related to provision of services. Focused on efficiency, monetary gain must be a central concern in their decision-making. Unfortunately, studies examining the link between digital communication and digital performance do not account for these concerns. For this reason we have introduced three dimensions that need to be taken into consideration, namely: type of stakeholder; performance effectiveness and efficiency; and individual and organizational levels of targeting.

This taxonomy enables a clearer view of the basic links that can be established, because "targeting" in digital communication via the Internet is essential for getting benefits related to attainment of effectiveness or efficiency-oriented goals. For example, it may be easier to get donations from organizations supporting the same goals. Private organizations such as philanthropic institutes are often more aware of nonprofit practices. They may be less willing to donate via the net, but are more likely to support goals indirectly. Similarly, public agencies supporting the provision of services are probably less interested in DC since their activities are not autonomous but reflect public policies and allocation of funds.

DC oriented towards public institutions will attain higher effectiveness, i.e. moral support and recognition of organizational goals in both service and advocacy groups. In many ways, this type of DC is more capable of raising public awareness, shaping organizational prestige, and increasing exposure to immediate and long-range goals of nonprofits. In such cases, expectations of increased efficiency, i.e. more donations and willingness to participate in organizational activities will be lower. However, when DC aims at higher personal contact with either individual or institutional agents, it will be more successful in attaining donations and contributions of all kinds.

It should be noted that some degree of effectiveness and efficiency is only possible when DC is highly sophisticated and costly. Some nonprofits, usually large international social and complex enterprises, will choose DC purely for effectiveness, though the impact of extended DC will probably have positive outcomes for efficient operation as well. For example, organizations such as Amnesty, Doctors of the World, Unique, Oxfam etc. have achieved high levels of contributions from effective use of the DC link. We therefore recommend using DC in the nonprofits, bearing in mind that direct involvement in attainment of a social goal is the rationale for participation. Providing "passive" support to a cause does not motivate individuals. The Internet certainly plays an important role, but it is essentially by physical presence that nonprofits are able to cope with the immediate and efficient achievement of social goals.

4. References

Ackcin, D. 2001. Non-profit marketing: just how far has it come? *Non-profit World, 19, 1,* 33-35.

Balser, D. & J. McCLusky. 2005. Managing stakeholder relationships and nonprofit organizational effectiveness, *Nonprofit Management and Leadership,* 15(3), 295-315.

Banaji, S. & Buckingham, D. 2009. The civic sell- Young people, the internet, and ethical consumption .*Communication & Society,* 12: 8, 1197-1223.

Barraket, J. 2005. Online opportunities for civic engagement? An examination of australian third sector organisations on the Internet .*Australian Journal of Emerging Technologies and Society* Vol. 3, No. 1.pp: 17-30

Baruch, Y. & N. Ramalho 2006. Communalities and distinctions in the measurement of organisational performance and effectiveness across for-profit and non-profit sectors, *Non-profit & Voluntary Sector Quarterly. 53,* 456-501.

Bielefeld, W. 1992. Uncertainty and non-profit strategies in the 1980's, *Nonprofit Management and Leadership, 2*, 4, 381-401.

Crittenden, W. F. 2000. Spinning straw into gold: The tenuous strategy, funding, and financial performance linkage, *Non-profit and Voluntary Sector Quarterly, 29*, 164-182.

Galaskiewicz, J. & Bielefeld. W. 1998. *Non-profit organisations in an age of uncertainty: A study of organisational change* New York: Aldine De Gruyter.

Hamburger, Y.A. 2008. Potential and promise of online volunteering *Computers in Human Behavior, 24*, 544–562

Hargittai, E. 2000. standing before the portals: non-profit websites in an age of commercial gatekeepers. *Sociology Dept*, vol. 2, no. 6, pp. 543-550.

Hart, T. R. 2002. E-Philanthropy: Using the Internet to build support' *International Journal of Non-profit and Voluntary Sector Marketing, 7*, No. 4, pp. 353 – 360.

Herman, R.D. & Renz, D.O. 2008. Advancing nonprofit organizational effectiveness research and theory: nine theses. *Nonprofit Management and Leadership, 18*, 4, 399-415.

Jennings, M. K & Zeitner., V. 2003. Internet use and civic engagement: a longitudinal analysis. *The Public Opinion Quarterly, 67*, No. 3 . pp. 311-334.

Keith, R. S., Arthur, G. E. & Michael, B. H. (1997). The contribution of local media to community involvement. *Journalism and Mass Communication Quarterly; 74*, 1 .pg. 97-107.

Mano, R. 2009. Information technology, adaptation and innovation in nonprofit human service organizations, *Technology in Human Services, 27* (3): 227-234.

Mano, R. (2010). Organizational crisis, adaptation, and innovation in Israel's nonprofit organizations: a learning approach, *Administration in Social Work, 34*(4), 344-350.

Mesch, S. G. & I. Talmud (2010). Internet connectivity, community participation, and place attachment: a longitudinal study *American Behavioral Scientist 53*: 45-65.

Olsen, M. Keevers, M. L., Paul, J. & Covington, S. 2001. E-relationship development strategy for the non-profit fundraising professional', *International Journal of Non-profit and Voluntary Sector* Marketing, Vol. 6, No. 4, pp.–73.

Postigo, H. 2009. America online volunteers: Lessons from an early co-production community *International Journal of Cultural Studies 12*: 451-469.

Schmid, H. 2002. Relationships between organisational properties and organisational effectiveness in three types of non-profit human service organisations, *Public Personnel Management, 31*, 377-395.

Shah, D. V. Nojin, K. &. Lance, H.R. 2001. "Connecting" and "Disconnecting" with civic life: patterns of internet use and the production of social capital, *Political Communication*, 18: 2, 141 – 162

Shah, D.V. Cho, J. Eveland, W.P. & Kwak, N. 2005. Information and expression in a digital age: modeling internet effects on civic participation. *Communication Research 32*: 531-565.

Shah, D. V. Mcleod, J. & Yoon, S.H. 2001. Communication, context, and community: an exploration of print, broadcast, and internet influences *Communication Research 28*: 464-506

Suárez, D.F. 2009. Nonprofit advocacy and civic engagement on the Internet *Administration & Society*. 41: 267 -289.

Tapia, A.H. & Ortiz, J.A. 2010. Network hopes: municipalities deploying wireless internet to increase civic engagement. *Social Science Computer Review* 28: 93-117

3

MANET Routing Protocols Performance Evaluation in Mobility

C. Palanisamy and B. Renuka Devi

Department of Information Technology
Bannari Amman Institute of Technology
Sathyamangalam, Tamil Nadu
India

1. Introduction

Ad-hoc network is a collection of wireless mobile nodes dynamically forming a temporary network without the aid of any established infrastructure or centralized administration. Routing protocols in mobile ad-hoc network helps node to send and receive packets. In this chapter our focus is to study Reactive (AODV and A-AODV), Proactive (DSDV) based on Random Waypoint mobility model. In this chapter, a new routing protocol A-AODV is proposed. The performance of three routing protocols (AODV, DSDV, and A-AODV) based on metrics such as packet delivery ratio, end to end delay, and throughput are studied. The simulation work is done with the NS-2.34 simulator with the ordered traffic load.

Mobile ad-hoc networks, also known as short-lived networks, are autonomous systems of mobile nodes forming network in the absence of any Centralized support. This is a new form of network and might be able to provide services at places where it is not possible otherwise. Absence of fixed infrastructure poses several types of challenges for this type of networking. Among these challenges is routing, which is responsible to deliver packets efficiently to mobile nodes. So routing in mobile ad-hoc network is a challenging task due to node mobility. Moreover bandwidth, energy and physical security are limited. MANET is the Art of Networking without a Network. In the recent years communication technology and services have advanced. Mobility has become very important, as people want to communicate anytime from and to anywhere. In the areas where there is little or no infrastructure is available or the existing wireless infrastructure is expensive and inconvenient to use, Mobile Ad hoc NETworks, called MANETs, are useful. They form the integral part of next generation mobile services. A MANET is a collection of wireless nodes that can dynamically form a network to exchange information without using any pre-existing fixed network infrastructure.

Each device in a MANET is free to move independently in any direction, and will therefore change its links to other devices frequently. Each must forward traffic unrelated to its own use, and therefore be a router. The primary challenge in building a MANET is equipping each device to continuously maintain the information required to properly route traffic. Such networks may operate by themselves or may be connected to the larger Internet.

2. Challenges in MANETs

The major open Challenges (S.Corson & J.Macker, 1999; H.Yang et al., 2004) to MANETs are listed below:

a. **Autonomous-** No centralized administration entity is available to manage the operation of different mobile nodes.

b. **Dynamic topology-** Nodes are mobile and can be connected dynamically in an arbitrary manner. Links of the network vary timely and are based on the proximity of one node to another node.

c. **Device discovery-** Identifying relevant newly moved in nodes and informing about their existence need dynamic update to facilitate automatic optimal route selection.

d. **Bandwidth optimization-** Wireless links have significantly lower capacity than the wired links.

e. **Limited resources-** Mobile nodes rely on battery power, which is a scarce resource. Also storage capacity and power are severely limited.

f. **Scalability-** Scalability can be broadly defined as whether the network is able to provide an acceptable level of service even in the presence of a large number of nodes.

g. **Limited physical security-** Mobility implies higher security risks (H.Yang, et al., 2004) such as peer-to- peer network architecture or a shared wireless medium accessible to both legitimate network users and malicious attackers. Eavesdropping, spoofing and denial-of-service attacks should be considered.

h. **Infrastructure-** less and self operated- Self healing feature demands MANET should realign itself to blanket any node moving out of its range.

i. **Poor Transmission Quality-** This is an inherent problem of wireless communication caused by several error sources that result in degradation of the received signal.

j. **Adhoc addressing-** Challenges in standard addressing scheme to be implemented.

k. **Network configuration-** The whole MANET infrastructure is dynamic and is the reason for dynamic connection and disconnection of the variable links.

l. **Topology maintenance-** Updating information of dynamic links among nodes in MANETs is a major challenge.

Wireless application scenarios lead to a diverse set of service requirements for the future Internet as summarized below:

1. Naming and addressing flexibility.
2. Mobility support for dynamic migration of end-users and network devices.
3. Location services that provide information on geographic position.
4. Self-organization and discovery for distributed control of network topology.
5. Security and privacy considerations for mobile nodes and open wireless channels.
6. Decentralized management for remote monitoring and control.
7. Cross-layer support for optimization of protocol performance.
8. Sensor network features such as aggregation, content routing and in-network Processing.
9. Cognitive radio support for networks with physical layer adaptation.
10. Economic incentives to encourage efficient sharing of resources.

3. Overview of AODV, DSDV and RWMM

3.1 Ad Hoc on-demand distance vector routing (AODV)

The AODV protocol (Perkins & Royer, 1999) is a reactive routing protocol for MANETs. It discovers routes once demanded via a route discovery process. The protocol uses route request (RREQ) packets sent by the sender node and circulating throughout the network. Each node in the network rebroadcasts the message except the sink node. The receiver replies to the RREQ message with a route reply (RREP) packet that is routed back to the sender node. The route is then cached for future reference; however in case a link is broken, a route error (RERR) packet is sent to the sender and to all the nodes as well so as to initiate a new route discovery. To maintain routing information, AODV uses a routing table with one entry for each destination. Thus, the table is used to propagate RREP to the source node. AOOV (Geetha Jayakumar & G. Gopinath, 2007, 2008) also relies on time, which means that if a routing table is not used recently, it will expire. Moreover, once a RERR is sent, it is meant to warn all nodes in the network; hence, this makes it very efficient to detect broken paths.

3.2 Destination-sequenced distance vector routing (DSDV)

The DSDV protocol (C. Perkins & P.Bhagwat, 1994) is a routing algorithm that focuses on finding the shortest paths. The protocol is based on the Bellman-Ford algorithm to find the routes with improvements. The latter algorithm is very similar to the well-known Dijkstra's algorithm with the support of negative weights. DSDV (Md. Monzur Morshed et al., 2010) falls in the proactive category of routing protocols; hence, every mobile node maintains a table containing all the available destinations, the number of hops to reach each destination, and a sequence number. The sequence number is assigned by the destination node its purpose is to distinguish between old nodes and new ones. In order for the nodes to keep track of moving other nodes, a periodic message containing a routing table is sent by each node to its neighbors. The same message can also be sent if significant change occurs at the level of the routing table. Therefore, the update of the routing table is both time-driven and event-driven. Further discussion can be done for better performance, such as not sending the whole table (full dump update), but only the modified portions (incremental update). The motivation behind it is to be able to update the rest of the network through one packet. This means that if the update requires more than one packet, a full dump is probably a safer approach in this case.

3.3 Random waypoint mobility model (RWMM)

Random Waypoint (RWP) model is a commonly used synthetic model for mobility, e.g., in Ad Hoc networks. It is an elementary model which describes the movement pattern of independent nodes by simple terms.

The Random Waypoint model (F. Bai et al., 2003; Guolong Lin et al., 2004; F.Bai & A.Helmy, 2004) is the most commonly used mobility model in research community. At every instant, a node randomly chooses a destination and moves towards it with a velocity chosen randomly from a uniform distribution [0, V_{max}], where V_{max} is the maximum allowable velocity for every mobile node. After reaching the destination, the node stops for a duration defined by the 'pause time' parameter. After this duration, it again chooses a random destination and repeats the whole process until the simulation ends.

4. Altered ad-hoc on-demand distance vector (A-AODV)

Analyzing previous protocols, we can say that most of on-demand routing protocols, except multipath routing, uses single route reply along the first reverse path to establish routing path. In high mobility, pre-decided reverse path can be disconnected and route reply message from destination to source can be missed. In this case, source node needs to retransmit route request message. Purpose of this study is to increase possibility of establishing routing path with less RREQ messages than other protocols have on topology change by nodes mobility. Specifically, the proposed A-AODV protocol discovers routes on-demand using a reverse route discovery procedure. During route discovery procedure, source node and destination node plays same role from the point of sending control messages. Thus after receiving RREQ message, destination node floods reverse request (R-RREQ), to find source node. When source node receives an R-RREQ message, data packet transmission is started immediately.

4.1 Route discovery

Since A-AODV is reactive routing protocol, no permanent routes are stored in nodes. The source node initiates route discovery procedure by broadcasting. The RREQ message contains information such as: message type, source address, destination address, broadcast ID, hop count, source sequence number destination sequence number, request time (timestamp). Whenever the source node issues a new RREQ, the broadcast ID is incremented by one. Thus, the source and destination addresses, together with the broadcast ID, uniquely identify this RREQ packet. The source node broadcasts the RREQ to all nodes within its transmission range. These neighboring nodes will then pass on the RREQ to other nodes in the same manner. As the RREQ is broadcasted in the whole network, some nodes may receive several copies of the same RREQ. When an intermediate node receives a RREQ, the node checks if already received a RREQ with the same broadcast id and source address. The node cashes broadcast id and source address for first time and drops redundant RREQ messages. The procedure is the same with the RREQ of AODV. When the destination node receives first route request message, it generates so called reverse request (R-RREQ) message and broadcasts it to neighbor nodes within transmission range like the RREQ of source node does. R-RREQ message contains the information such as: reply source id, reply destination id, reply broadcast id, hop count, destination sequence number, reply time (timestamp). When broadcasted R-RREQ message arrives to intermediate node, it will check for redundancy. If it already received the same message, the message is dropped, otherwise forwards to next nodes. Furthermore, node stores or updates following information of routing table:

- Destination and Source Node Address
- Hops up to destination
- Destination Sequence Number
- Route expiration time and next hop to destination node.

And whenever the original source node receives first R-RREQ message it starts packet transmission, and late arrived R-RREQs are saved for future use. The alternative paths can be used when the primary path fails communications.

4.2 Route update and maintenance

When control packets are received, the source node chooses the best path to update, i.e. first the node compares sequence numbers, and higher sequence numbers mean recent routes. If sequence numbers are same, then compares number of hops up to destination, routing path with fewer hops is selected. Since the wireless channel quality is time varying, the best path varies over time. The feedback from the MAC layer can be used to detect the connectivity of the link. When a node notifies that its downstream node is out of its transmission range, the node generates a route error (RERR) to its upstream node. If fail occurs closer to destination node, RRER received nodes can try local-repair, otherwise the nodes forward RRER until it reaches the source node. The source node can select alternative route or trigger a new route discovery procedure. There will be at least a single path for route reply so throughput will be increased although there is high mobility.

5. Performance metrics

There are various performance metrics. As suggested (I. Awan & K. Al-Begain, 2006; V.Ramesh et al., 2010) packet delivery fraction and end to end delay is considered as two basic performance metrics. Also (S.H Manjula et al., 2008) suggested to use random way point mobility model for considering the mobility pattern of nodes in my simulation. In terms of delay and dropped packet (Rachid Ennaji & Mohammed Boulmalf, 2009) performance of AODV and DSDV were measured.

5.1.1 Packet delivery fraction

It is calculated by dividing the number of packets received by the destination through the number of packets originated by the application layer of the source. It specifies the packet loss rate, which limits the maximum throughput of the network.

5.1.2 End-to-end delay

It is the average time it takes a data packet to reach the destination. This metric is calculated by subtracting time at which first packet was transmitted by source from time at which first data packet arrived to destination. This metric is significant in understanding the delay introduced by path discovery.

5.1.3 Throughput

Throughput refers to how much data can be transferred from one location to another in a given amount of time.

5.2 Test scenarios

The simulation is conducted in three different scenarios. In the first scenario, the comparison of the three routing protocols is compared in various numbers of nodes. The number of nodes is set to 10, 20, 30, 40 and 50 while the simulation time is fixed. In the second scenario, the routing protocols are evaluated in different simulation time while the number of nodes is fixed. The number of nodes is set to 30. The simulation time are set to 175, 225, 400, 575 and 750 second. In the third scenario, the routing protocols are evaluated in different node speed. The node speed is set to 10, 20, 30, 40 and 50 m/s. The number of nodes is fixed to 30. Random Waypoint Mobility Model in common to the three scenarios considered below.

5.2.1 Test scenario 1: Varying numbers of nodes

In this scenario, all the three routing protocol are evaluated based on the three performance metric which are Packet Delivery Fraction, End-to-End Delay and Throughput. The simulation environments for this scenario are:

- Various numbers of nodes are 10, 20, 30, 40 and 50.
- Simulation Time is set to 175 seconds
- Area size is set to 500 x 500
- Node Speed is random varies between 5 to 12 m/s

5.2.1.1 Packet delivery fraction

In the figure 1, x- axis represents the varying number of nodes and y- axis represents the packet delivery fraction. Figure 1, shows that A-AODV perform better when the number of nodes increases because nodes become more stationary will lead to more stable path from source to destination. DSDV and AODV performance dropped as number of nodes increase because more packets dropped due to link breaks. DSDV is better than AODV when the number of nodes increases. A-AODV improved the PDF since it has a definite route to destination without any link break.

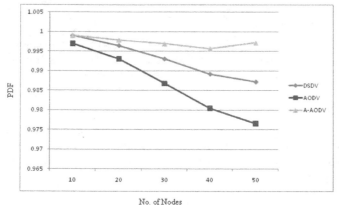

No. of Nodes

Fig. 1. Packet Delivery Fraction in Scenario 1

5.2.1.2 End-to-end delay

In the figure 2 x- axis represents the varying number of nodes and y- axis represents the end to end delay in mille seconds. A-AODV does not produce so much delay even the number of nodes increased. It is better than the other two protocols. The performance of DSDV is slightly better than AODV especially when the number of nodes cross 30. It shows that, the DSDV protocol has greater delay than AODV.

This is mainly because of the stable routing table maintenance. A-AODV produces lower delay due to the fact that it uses flooding scheme in the route reply. Thus the delay is reduced to a greater extent.

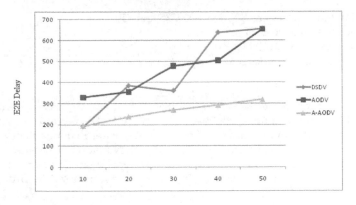

No of Nodes

Fig. 2. End-to-End Delay in Scenario 1

5.2.1.3 Throughput

In the figure 3, x- axis represents the varying number of nodes and y- axis represents the throughput.

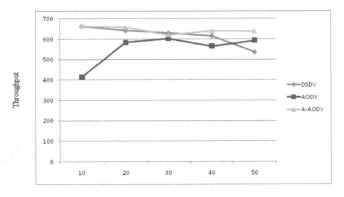

No of Nodes

Fig. 3. Throughput in Scenario 1

From Figure 3, it is observed that DSDV is less prone to route stability compared to AODV when number of nodes increased. For A-AODV, the route stability is more so the throughput does not varied when number of nodes increases. DSDV protocol produces less throughputs when number of nodes are increased especially with a marginal difference after number of nodes is increased to 40.

5.2.2 Test scenario 2: Varying simulation time

In this scenario, all the three routing protocol are evaluated based on the three performance metric which are Packet Delivery Fraction, End-to-End Delay and Throughput. The simulation environments for this scenario are:

- Various simulation times are 175, 225, 400, 575 and 750 seconds
- Number of nodes is fixed to 30
- Area size is set to 1000 x 1000
- Node Speed is random varies between 5 to 12 m/s

5.2.2.1 Packet delivery fraction

In the figure 4, x- axis represents the varying simulation time and y- axis represents the packet delivery fraction. Based on Figure 4, contrast to AODV, A-AODV performs better. It delivers the data packet regardless to mobility rate. DSDV has the better PDF rate than AODV which has the great variation in packet drop in 225.This great variation is because of more link failures due to mobility. For AODV, it shows significant dependence on route stability, thus its PDF is lower when the nodes change its position as simulation time increased.

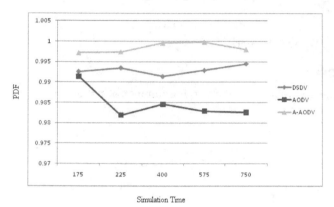

Simulation Time

Fig. 4. Packet Delivery Fraction in Scenario 2

5.2.2.2 End-to-end delay

In the figure 5, x- axis represents the varying simulation time and y- axis represents the end to end delay in mille seconds. From the figure 5, it is inferred that A-AODV exhibits lower

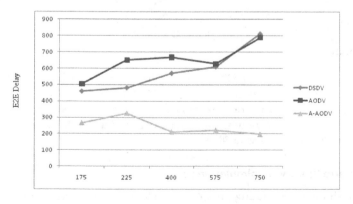

Simulation Time

Fig. 5. End-to-End Delay in Scenario 2

average end-to-end delay all the time regardless to node mobility rate compared to the other two protocols. AODV uses a flooding scheme in route reply to create a definite route to destination to avoid link breaks. So it has lower end-to-end delay time compare to others. AODV and DSDV exhibit more or less same end-to-end delay.

5.2.2.3 Throughput

In the figure 6, x- axis represents the varying simulation time and y- axis represents the throughput. A-AODV produces better results on Throughput than the other two protocols. This is due to the route maintenance. The other two protocols fluctuate when simulation time increases because of the instability in their routing paths and link failures.

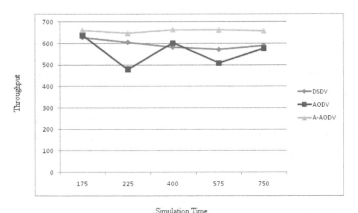

Fig. 6. Throughput in Scenario 2

5.2.3 Test scenario 3: Varying node speed

In this scenario, all the three routing protocol are evaluated based on the three performance metric which are Packet Delivery Fraction, End-to-End Delay and Throughput. The simulation environments for this scenario are:

- Various node speeds are 10, 20, 30, 40 and 50 m/s
- Simulation time is 175 seconds
- Number of nodes is fixed to 30
- Area size is set to 1000 x 1000.

5.2.3.1 Packet delivery fraction

In the figure 7, x - axis represents the varying node speed and y- axis represents the packet delivery fraction. From figure 7, it is shown that the speed of the node has less impact on DSDV protocol when node speed is up to 40. The PDF losses after node speed 40 because of the link breakage due to mobility. The node speed does not affect the PDF of the protocols AODV and A-AODV. Generally the PDF decreases in AODV protocol than other two protocols because the data transfer process need a new route discovery due to mobility. The PDF is increased in A-AODV protocol because of the multiple route-reply scheme.

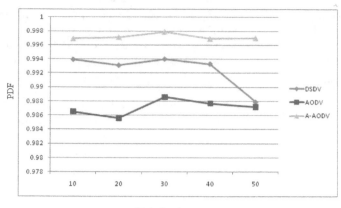

Node Speed

Fig. 7. Packet Delivery Fraction in Scenario 3

5.2.3.2 End-to-end delay

In the figure 8, x- axis represents the varying node speed and y- axis represents the end to end delay in mille seconds. Based on Figure 8, for varying speed, A-AODV produces less End to End Delay, but the performance of DSDV is slightly better than AODV.

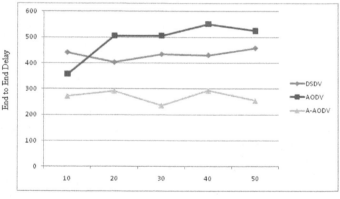

Node Speed

Fig. 8. End-to-End Delay in Scenario 3

The End-to End Delay is lowed in DSDV than AODV because of the proactive nature of the protocol. While considering End-to-End Delay in various scenarios A-AODV protocol works better than other two protocols because of the flooding scheme in route reply. The flooding scheme with broadcast ID in the route reply make the delay lower than the other two protocols.

5.2.3.3 Throughput

In the figure 9, x- axis represents the varying node speed and y- axis represents the throughput.

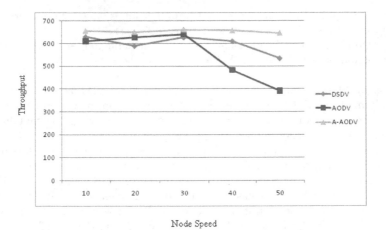

Node Speed

Fig. 9. Throughput in Scenario 3

The performance of A-AODV is almost same in various speeds. The throughput is maintained for various node speeds. AODV protocol has the continuous throughput decrease as node speed increases after the node speed 30 because the route table cannot be maintained to the speed of the node. In the case of DSDV protocol the proactive nature of the protocol produces the decreased throughput as node speed increases.

6. Conclusion

The performance of all the routing protocol are measured with respect to metrics namely, Packet Delivery Fraction, End to End Delay and Throughput in three different scenarios: simulation time, number of nodes and node speed. The results indicate that the performance of A-AODV is superior to regular AODV and DSDV. It is also observed that the performance is improved, especially when the number of nodes in the network is increased. When the number of nodes is increased beyond 30 and above, the performance of all the three protocols varies very much. It is due to the fact that, lot of control packets are generated in the network. It is also observed that A-AODV is even better than DSDV protocol in PDF, lower end to end delay and throughput. DSDV is better than AODV protocol in PDF. It is concluded that A-AODV improved the PDF and lowered end-to-end delay when the number of nodes are increased. Throughput is also improved. AODV has lower performance compared to DSDV in most of the scenarios.

7. References

S. Corson and J. Macker, Mobile Ad hoc Networking (MANET): Routing Protocol Performance Issues and Evaluation Considerations, *RFC: 2501*, January 1999.

C.E. Perkins and E.M. Royer, Ad-Hoc on-demand distance Vector Routing, *Proceedings of 2nd IEEE Workshop on Mobile Computing and Application, WMCSA '99*, pp. 90-100, 25-26 Feb 1999.

Charles E. Perkins, Pravin Bhagwat, 'Highly dynamic Destination-Sequenced Distance-Vector routing (DSDV) for mobile computers", SIGCOMM Comput. Commun. Rev. In

Proceedings of the conference on Communications architectures, protocols and applications, Vol. 24, No. 4. October 1994), pp. 234-244

Md. Monzur Morshed, Md. Habibur Rahman, Md.Reazur Rahman Manzumder, Performance evaluation of DSDV and AODV Routing Protocols in Mobile Ad-hoc Networks, 4*th* *International conference on new trends in information science and service science (NISS)*, On page(s): 399-403 Location: Gyeongju Print ISBN: 978-1-4244-6982-6

Geetha Jayakumar and G. Gopinath, Performance comparison of two on-demand routing protocols for Ad-hoc networks based on random waypoint mobility model, *American Journal of Applied Sciences*, pp. 659-664, June 2008

Geetha Jayakumar and Gopinath Ganapathy, "Performance comparison of Mobile Ad-hoc Network Routing Protocol," International Journal of Computer Science and Network Security (JJCSNS 2007), vol. VII, no. I I, pp. 77-84, November 2007

V. Ramesh, P. Subbaiah, N. Koteswar Rao and M. Janardhana Raju, Performance comparison and analysis of DSDV and AODV for MANET, (JJCSE) *International Journal on Computer Science and Engineering* , vol. 02 , pp. 183-188, 2010

S. H. Manjula, C. N. Abhilash, Shaila K., K. R. Venugopal, L. M. Patnaik, Performance of AODV Routing Protocol using group and entity Mobility Models in Wireless Sensor Networks, *Proceedings of the International Multi Conference of Engineers and Computer Scientists (IMECS 2008)*, vol. 2, 19-21 March 2008, Hong Kong, pp. 1212-1217.

Rachid Ennaji, Mohammed Boulmalf, Routing in wireless sensor networks, in *International conference on multimedia computing and systems, ICMCS '09* , page iv-x ISBN:978-1-4244-3756-6, DOI:10.1109/MMCS.2009.5256626.

I. Awan and K. Al-Begain, Performance evaluation of wireless networks, *International Journal of Wireless Information Networks*, vol. 13, no. 2, pp. 95–97, 2006.

J. Broch, D. Johnson, and D. Maltz, The dynamic source routing protocol for mobile ad hoc networks, *IETF Internet Draft*, December 1998, http://www.ietf.org/internet-drafts/

A. E. Mahmoud, R. Khalaf & A, Kayssi, Performance Comparison of the AODV and DSDV Routing Protocols in Mobile Ad-Hoc Networks, Lebanon, 2007.

F. Bai, N. Sadagopan, and A. Helmy, Important: a framework to systematically analyze the impact of mobility on performance of routing protocols for ad hoc networks, in *Proceedings of IEEE Information Communications Conference (INFOCOM 2003)*, SanFrancisco, Apr. 2003.

H. Yang, H. Luo, F. Ye, S. Lu, and L. Zhang, Security in Mobile Ad Hoc Networks: Challenges and Solutions, *IEEE Wireless Communications*, pp. 38-47, 2004.

E. M. Royer and C.K.Toh. A Review of Current Routing Protocol for Ad-Hoc Mobile Wireless Networks, *IEEE Personal Communications*. pp. 46-55. April 1999.

F. Bai, A. Helmy, A Survey of Mobility Modelling and analysis in wireless Adhoc Networks in *Wireless Ad Hoc and Sensor Networks*, Kluwer Academic Publishers, 2004.

Guolong Lin, Guevara Noubir and Rajmohan Rajaraman, Mobility Models for Ad Hoc Network Simulation, in *Proceedings of IEEE INFOCOM* 2004.

The NS-2 tutorial homepage *http://www.isi.edu/nsnam/ns/tutorial/index.html* volume 1 pp 7-11, 2004.

4

Adaptive Blind Channel Equalization

Shafayat Abrar[1], Azzedine Zerguine[2] and Asoke Kumar Nandi[3]
[1]*Department of Electrical Engineering,*
COMSATS Institute of Information Technology, Islamabad 44000
[2]*Department of Electrical Engineering,*
King Fahd University of Petroleum and Minerals, Dhahran 31261
[3]*Department of Electrical Engineering and Electronics,*
The University of Liverpool, Liverpool L69 3BX
[1]*Pakistan*
[2]*Saudi Arabia*
[3]*United Kingdom*

1. Introduction

For bandwidth-efficient communication systems, operating in high inter-symbol interference (ISI) environments, adaptive equalizers have become a necessary component of the receiver architecture. An accurate estimate of the amplitude and phase distortion introduced by the channel is essential to achieve high data rates with low error probabilities. An adaptive equalizer provides a simple practical device capable of both learning and inverting the distorting effects of the channel. In conventional equalizers, the filter tap weights are initially set using a training sequence of data symbols known both to the transmitter and receiver. These trained equalizers are effective and widely used. Conventional least mean square (LMS) adaptive filters are usually employed in such supervised receivers, see Haykin (1996).

However, there are several drawbacks to the use of training sequences. Implementing a training sequence can involve significant transceiver complexity. Like in a point-to-multipoint network transmissions, sending training sequences is either impractical or very costly in terms of data throughput. Also for slowly varying channels, an initial training phase may be tolerable. However, there are scenarios where training may not be feasible, for example, in equalizer implementations of digital cellular handsets. When the communications environment is highly non-stationary, it may even become grossly impractical to use training sequences. A blind equalizer, on the other hand, does not require a training sequence to be sent for start-up or restart. Rather, the blind equalization algorithms use a priori knowledge regarding the statistics of the transmitted data sequence as opposed to an exact set of symbols known both to the transmitter and receiver. In addition, the specifications of training sequences are often left ambiguous in standards bodies, leading to vendor specific training sequences and inter-operability problems. Blind equalization solves this problem as well, see Ding & Li (2001); Garth et al. (1998); Haykin (1994).

In this Chapter, we provide an introduction to the basics of adaptive blind equalization. We describe popular methodologies and criteria for designing adaptive algorithms for blind equalization. Most importantly, we discuss how to use the probability density function (PDF) of transmitted signal to design ISI-sensitive cost functions. We discuss the issues of admissibility of proposed cost function and stability of derived adaptive algorithm.

2. Trained and blind adaptive equalizers: Historical perspectives

Adaptive trained channel equalization was first developed by Lucky for telephone channels, see Lucky (1965; 1966). Lucky proposed the so-called zero-forcing (ZF) method to be applied in FIR equalization. It was an adaptive procedure and in a noiseless situation, the optimal ZF equalizer tends to be the inverse of the channel. In the mean time, Widrow and Hoff introduced the least mean square (LMS) adaptive algorithm which begins adaptation with the aid of a training sequence known to both transmitter and receiver, see Widrow & Hoff (1960); Widrow et al. (1975). The LMS algorithm is capable of reducing mean square error (MSE) between the equalizer output and the training sequence. Once the signal eye is open, the equalizer is then switched to tracking mode which is commonly known as decision-directed mode. The decision-directed method is unsupervised and its effectiveness depends on the initial condition of equalizer coefficients; if the initial eye is closed then it is likely to diverge.

In blind equalization, the desired signal is unknown to the receiver, except for its probabilistic or statistical properties over some known alphabets. As both the channel and its input are unknown, the objective of blind equalization is to recover the unknown input sequence based solely on its probabilistic and statistical properties, see C.R. Johnson, Jr. et al. (1998); Ding & Li (2001); Haykin (1994). Historically, the possibility of blind equalization was first discussed in Allen & Mazo (1974), where the authors proved analytically that an adjusting equalizer, optimizing the mean-squared sample values at its output while keeping a particular tap anchored at unit value, is capable of inverting the channel without needing a training sequence. In the subsequent year, Sato was the first who came up with a robust realization of an adaptive blind equalizer for PAM signals, see Sato (1975). It was followed by a number of successful attempts on blind magnitude equalization (i.e., equalization without carrier-phase recovery) in Godard (1980) for complex-valued signals (V29/QPSK/QAM), in Treichler & Agee (1983) for AM/FM signals, in Serra & Esteves (1984) and Bellini (1986) for PAM signals. However, many of these algorithms originated from intuitive starting points.

The earliest works on joint blind equalization and carrier-phase recovery were reported in Benveniste & Goursat (1984); Kennedy & Ding (1992); Picchi & Prati (1987); Wesolowski (1987). Recent references include Abrar & Nandi (2010a;b;c;d); Abrar & Shah (2006a); Abrar & Qureshi (2006b); Abrar et al. (2005); Goupil & Palicot (2007); Im et al. (2001); Yang et al. (2002); Yuan & Lin (2010); Yuan & Tsai (2005). All of these blind equalizers are capable of recovering the true power of transmitted data upon convergence and are classified as *Bussgang*-type, see Bellini (1986). The Bussgang blind equalization algorithms make use of a nonlinear estimate of the channel input. The memoryless nonlinearity, which is the function of equalizer output, is designed to minimize an ISI-sensitive cost function that implicitly exploits higher-order statistics. The performance of such kind of blind equalizer depends strongly on the choice of nonlinearity.

The first comprehensive analytical study of the blind equalization problem was presented by Benveniste, Goursat, and Ruget in Benveniste et al. (1980a;b). They established that if the transmitted signal is composed of non-Gaussian, independent and identically distributed samples, both channel and equalizer are linear time-invariant filters, noise is negligible, and the probability density functions of transmitted and equalized signals are equal, then the channel has been perfectly equalized. This mathematical result is very important since it establishes the possibility of obtaining an equalizer with the sole aid of signal's statistical properties and without requiring any knowledge of the channel impulse response or training data sequence. Note that the very term "blind equalization" can be attributed to Benveniste and Goursat from the title of their paper Benveniste & Goursat (1984). This seminal paper established the connection between the task of blind equalization and the use of higher-order statistics of the channel output. Through rigorous analysis, they generalized the original Sato algorithm into a class of algorithms based on non-MSE cost functions. More importantly, the convergence properties of the proposed algorithms were carefully investigated.

The second analytical landmark occurred in 1990 when Shalvi and Weinstein significantly simplified the conditions for blind equalization, see Shalvi & Weinstein (1990). Before this work, it was usually believed that one needs to exploit infinite statistics to ensure zero-forcing equalization. Shalvi and Weinstein showed that the zero-forcing equalization can be achieved if only two statistics of the involved signals are restored. Actually, they proved that, if the fourth order cumulant (kurtosis) is maximized and the second order cumulant (energy) remains the same, then the equalized signal would be a scaled and rotated version of the transmitted signal. Interesting accounts on Shalvi and Weinstein criterion can be found in Tugnait et al. (1992) and Romano et al. (2011).

3. System model and "Bussgang" blind equalizer

The baseband model for a typical complex-valued data communication system consists of an unknown linear time-invariant channel $\{h\}$ which represents the physical inter-connection between the transmitter and the receiver. The transmitter generates a sequence of complex-valued random input data $\{a_n\}$, each element of which belongs to a complex alphabet $\mathcal{A}$. The data sequence $\{a_n\}$ is sent through the channel whose output x_n is observed by the receiver. The input/output relationship of the channel can be written as:

$$x_n = \sum_k a_{n-k} h_k + \bullet_n, \tag{1}$$

where the additive noise $\bullet_n$ is assumed to be stationary, Gaussian, and independent of the channel input a_n. We also assume that the channel is stationary, moving-average and has finite length. The function of equalizer at the receiver is to estimate the delayed version of original data, $a_{n-\bullet}$, from the received signal x_n. Let $w_n = [w_{n,0}, w_{n,1}, \cdots, w_{n,N-1}]^T$ be vector of equalizer coefficients with N elements (superscript T denotes transpose). Let $x_n = [x_n, x_{n-1} \cdots, x_{n-N+1}]^T$ be the vector of channel observations. The output of the equalizer is

$$y_n = w_n^H x_n \tag{2}$$

where superscript H denotes conjugate transpose. If $\{t\} = \{h\} \star \{w^*\}$ represents the overall channel-equalizer impulse response (where $\star$ denotes convolution), then (2) can be expressed as:

$$y_n = \sum_i w_i^* x_{n-i} = \sum_l a_{n-l} t_l + \bullet_n' = t \bullet a_{n-\bullet} + \underbrace{\sum_{l \neq \bullet} t_l a_{n-l} + \bullet_n'}_{\text{signal} + \text{ISI} + \text{noise}} \tag{3}$$

Equation (3) distinctly exposes the effects of multi-path inter-symbol interference and additive noise. Even in the absence of additive noise, the second term can be significant enough to cause an erroneous detection.

The idea behind a Bussgang blind equalizer is to minimize (or maximize), through the choice of equalizer filter coefficients w, a certain cost function J, depending on the equalizer output y_n, such that y_n provides an estimate of the source signal a_n up to some inherent indeterminacies, giving, $y_n = \bullet \, a_{n-\bullet}$, where $\bullet = |\bullet| e^{\bullet\bullet} \in \mathbb{C}$ represents an arbitrary gain. The phase $\bullet$ represents an isomorphic rotation of the symbol constellation and hence is dependent on the rotational symmetry of signal alphabets; for example, $\bullet = m \bullet /2$ radians, with $m \in \{0, 1, 2, 3\}$ for a quadrature amplitude modulation (QAM) system. Hence, a Bussgang blind equalizer tries to solve the following optimization problem:

$$w^\dagger = \arg \, \text{optimize}_w \, J, \, \text{with} \, J = \mathsf{E}[\mathcal{J}(y_n)] \tag{4}$$

The cost J is an expression for implicitly embedded higher-order statistics of y_n and $\mathcal{J}(y_n)$ is a real-valued function. Ideally, the cost J makes use of statistics which are significantly modified as the signal propagates through the channel, and the optimization of cost modifies the statistics of the signal at the equalizer output, aligning them with those at channel input. The equalization is accomplished when equalized sequence y_n acquires an identical distribution as that of the channel input a_n, see Benveniste et al. (1980a). If the implementation method is realized by stochastic gradient-based adaptive approach, then the updating algorithm is

$$w_{n+1} = w_n \pm \bullet \left(\frac{\bullet \mathcal{J}}{\bullet w_n} \right)^* \tag{5a}$$

$$= w_n + \bullet \, \Phi(y_n)^* x_n, \, \text{with} \, \Phi(y_n) = \pm \frac{\bullet \mathcal{J}}{\bullet y_n^*} \tag{5b}$$

where $\bullet$ is step-size, governing the speed of convergence and the level of steady-state performance, see Haykin (1996). The positive and negative signs in (5a) are respectively for maximization and minimization. The complex-valued error-function $\Phi(y_n)$ can be understood as an estimate of the difference between the desired and the actual equalizer outputs. That is, $\Phi(y_n) = \Psi(y_n) - y_n$, where $\Psi(y_n)$ is an estimate of the transmitted data a_n. The nonlinear memory-less estimate, $\Psi(y_n)$, is usually referred to as Bussgang nonlinearity and is selected such that, at steady state, when y_n is close to $a_{n-\bullet}$, the autocorrelation of y_n becomes equal to the cross-correlation between y_n and $\Psi(y_n)$, i.e.,

$$\mathsf{E}\left[y_n \Phi(y_{n-i})^*\right] = 0 \Rightarrow \mathsf{E}\left[y_n y_{n-i}^*\right] = \mathsf{E}\left[y_n \Psi(y_{n-i})^*\right]$$

An admissible estimate of $\Psi(y_n)$, however, is the conditional expectation $\mathsf{E}\left[a_{n'} | y_n\right]$, see Nikias & Petropulu (1993). Using Bayesian estimation technique, $\mathsf{E}\left[a_{n'} | y_n\right]$ was derived in

Bellini (1986); Fiori (2001); Haykin (1996); Pinchas & Bobrovsky (2006; 2007). These methods, however, rely on explicit computation of higher-order statistics and are not discussed here.

4. Trained and blind equalization design methodologies

Generally, a blind equalization algorithm attempts to invert the channel using both the received data samples and certain known statistical properties of the input data. For example, it is easy to show that for a minimum phase channel, the spectra of the input and output signals of the channel can be used to determine the channel impulse response. However, most communication channels do not possess minimum phase. To identify a non-minimum phase channel, a non-Gaussian signal is required along with nonlinear processing at the receiver using higher-order moments of the signal, see Benveniste et al. (1980a;b). Based upon available analysis, simulations, and experiments in the literature, it can be said that an admissible blind cost function has two main attributes: 1) it makes use of statistics which are significantly modified as the signal propagates through the channel, and 2) optimization of the cost function modifies the statistics of the signal at the channel output, aligning them with the statistics of the signal at the channel input.

Designing a blind equalization cost function has been lying strangely more in the realm of art than science; majority of the cost functions tend to be proposed on intuitive grounds and then validated. Due to this reason, a plethora of blind cost functions is available in literature. On the contrary, the fact is that there exist established methods which facilitate the designing of blind cost functions requiring statistical properties of transmitted and received signals. One of the earliest methods originated in late 70's in geophysics community who sought to determine the inverse of the channel in seismic data analysis and it was named minimum entropy deconvolution (MED), see Gray (1979b); Wiggins (1977). Later in early 90's, Satorius and Mulligan employed MED principle and came up with several proposals to blindly equalize the communication channels, see Satorius & Mulligan (1993). However, those marvelous signal-specific proposals regrettably failed to receive serious attention.

In the sequel, we discuss MED along with other popular methods for designing blind cost functions and corresponding adaptive equalizers.

4.1 Lucky criterion

In 1965, Lucky suggested that the propagation channel may be inverted by an equalizer if equalizer minimizes the following peak distortion criterion, see Lucky (1965):

$$J_{\text{peak}} = \frac{1}{|t_\bullet|} \sum_{l \neq \bullet} |t_l| \tag{6}$$

This criterion is equivalent to requiring that the equalizer maximizes the eye opening. the intuitive explanation of (6) is as follows. From (3), ignoring the noise, obtain the value of error $\mathcal{E}$ due to ISI, given as $\mathcal{E} = y_n - a_{n-\bullet} = a_{n-\bullet} \cdot (t_\bullet - 1) + \sum_{l \neq \bullet} t_l \, a_{n-l}$. Assuming the maximum and minimum values of a_n are $\tilde{a}$ and $-\tilde{a}$, respectively; the maximum error is easily written as

$$\mathcal{E}_{\text{max}} = |y_n - a_{n-\bullet}|_{\text{max}} = |a_{n-\bullet}| \, |t_\bullet - 1| + \tilde{a} \sum_{l \neq \bullet} |t_l| \tag{7}$$

If $t.$ is close to unity, then the cost J_{peak} is the scaled version of $\mathcal{E}_{max}$ as given by $J_{peak} \approx \mathcal{E}_{max}/(|t.|\bar{a})$. It is important to note that the cost J_{peak} is a convex function of the equalizer weights and has a well-defined minimum. Thus any minimum of J_{peak} found by gradient search or other systematic programming methods must be the absolute (or global) minimum of distortion, see Lucky (1966). To prove the convexity of J_{peak}, it is necessary to show that for two equalizer settings $w^{(a)}$ and $w^{(b)}$ and for all $\bullet$, $0 \leq \bullet \leq 1$

$$J_{peak}(\bullet\, w^{(a)} + (1 - \bullet)w^{(b)}) \leq \bullet\, J_{peak}(w^{(a)}) + (1 - \bullet)J_{peak}(w^{(b)}) \tag{8}$$

The above equation shows that the distortion always lies on or beneath the chord joining values of distortion in N-spaces. Below is the proof of (8):

$$J_{peak}(\bullet\, w^{(a)} + (1 - \bullet)w^{(b)}) = \sum_{l \neq \bullet} \left| \sum_k h_k(\bullet\, (w_{l-k}^{(a)})^* + (1 - \bullet)(w_{l-k}^{(b)})^*) \right|$$

$$\leq \bullet \sum_{l \neq \bullet} \left| \sum_k h_k(w_{l-k}^{(a)})^* \right| + (1 - \bullet) \sum_{l \neq \bullet} \left| \sum_k h_k(w_{l-k}^{(b)})^* \right| \tag{9}$$

$$= \bullet\, J_{peak}(w^{(a)}) + (1 - \bullet)J_{peak}(w^{(b)}).$$

However, in practice, it is not an easy one to achieve this convexity in a gradient search (adaptive) procedure, as it is necessary to obtain the projection of the gradient onto the constraint hyperplane, see Allen & Mazo (1973). Alternatively one can also seek to minimize the mean-square distortion criterion:

$$J_{ms} = \frac{1}{|t.|^2} \sum_{l \neq \bullet} |t_l|^2 \tag{10}$$

The cost J_{ms} is not convex but unimodal, mathematically tractable and capable of yielding admissible solution, see Allen & Mazo (1973). Under the assumption $|t.| = \max\{|t|\}$, Shalvi & Weinstein (1990) used the expression (10) to quantify ISI, i.e., ISI $= J_{ms}$. Using criterion (10), we can formulate the following tractable problem for ISI mitigation:

$$w^{\dagger} = \arg \min_w \sum_{l \neq \bullet} |t_l|^2 \quad \text{s.t.} \quad |t.| = 1. \tag{11}$$

where we have assumed that the equalizer coefficients $(w^{\dagger})$ have been selected such that the condition $(|t.| = 1)$ is always satisfied. Introducing the channel autocorrelation matrix $\mathcal{H}$, whose (i, j) element is given by $\mathcal{H}_{ij} = \sum_k h_{k-i}h_{k-j}^*$, we can show that $\sum_l |t_l|^2 = w_n^H \mathcal{H} w_n$. The equalizer has to make one of the coefficients of $\{t_l\}$ say $t. = t_0 = w_n^H h$ to be unity and others to be zero, where $h = [h_{K-1}, h_{K-2}, \cdots, h_1, h_0]^T$; it gives the value of ISI of an unequalized system at time index n as follows:

$$\text{ISI} = \frac{w_n^H \mathcal{H} w_n}{|w_n^H h|^2} - 1. \tag{12}$$

Now consider the optimization of the problem (11). Using Lagrangian multiplier • , we obtain

$$\sum_{l\neq\bullet} |t_l|^2 + \bullet\ (t_\bullet - 1) = w_n^H \mathcal{H} w_n - 1 + \bullet\ \left(w_n^H h - 1 \right) \tag{13}$$

Next differentiating with respect to w_n^* and equating to zero, we get $\mathcal{H} w_n + \bullet\ h = 0 \Rightarrow w_n = -\bullet\ \mathcal{H}^{-1} h$. Substituting the above value of w_n in (12), we obtain

$$ISI_n = \frac{h^H \mathcal{H}^{-1} h}{|h^H \mathcal{H}^{-1} h|^2} - 1. \tag{14}$$

To appreciate the possible benefit of solution (14), consider a channel $h_{-1} = 1 - \bullet$, $h_0 = \bullet$ and $h_1 = 0$, where $0 \leq \bullet \leq 1$. Without equalizer, we have

$$ISI = \frac{(1 - \bullet)^2 + \bullet^2}{\max(1 - \bullet, \bullet)^2} - 1. \tag{15}$$

The ISI approaches zero when • is either zero or unity. Assuming a 2-tap equalizer, we obtain

$$\mathcal{H} = \begin{bmatrix} (1 - \bullet)^2 + \bullet^2 & (1 - \bullet)\bullet \\ (1 - \bullet)\bullet & (1 - \bullet)^2 + \bullet^2 \end{bmatrix} \tag{16}$$

Using (14) and (16), we obtain

$$ISI = \frac{\bullet^2 (1 - \bullet)^2}{1 - 4\bullet + 6\bullet^2 - 4\bullet^3 + 2\bullet^4}. \tag{17}$$

Refer to Fig. 1, the ISI (17) of the equalized system is lower than that of the uncompensated system. The adaptive implementation of J_{ms} can be realized in a supervised scenario.

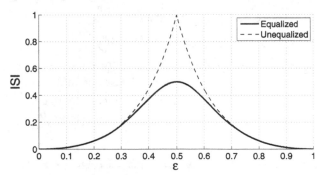

Fig. 1. ISI of unequalized and equalized systems.

Combining the two expression (7) and (10), the following cost is obtained which is usually termed as mean square error (MSE) criterion, see Widrow et al. (1975):

$$J_{mse} = E \left[|a_{n-\bullet} - y_n|^2 \right] \tag{18}$$

Minimizing (18), we obtain the following update:

$$w_{n+1} = w_n + \bullet \, (a_{n-\bullet} - y_n)^* x_n, \tag{19}$$

which is known as LMS algorithm or Widrow-Hoff algorithm. Note that the parameter $\bullet$ is a positive step-size and the following value of $\bullet \equiv \bullet_{LMS}$ ensures the stability of algorithm, see Farhang-Boroujeny (1998):

$$0 < \bullet_{LMS} < \frac{1}{2NP_a}, \tag{20}$$

where $P_a = \mathrm{E}\left[|a|^2\right]$ is the average energy of signal a_n and N is the length of equalizer.

4.2 Minimum entropy deconvolution criterion

The minimum entropy deconvolution (MED) is probably the earliest principle for designing cost functions for blind equalization. This principle was introduced by Wiggins in seismic data analysis in the year 1977, who sought to determine the inverse channel $w^\dagger$ that maximizes the *kurtosis* of the deconvolved data y_n, see Wiggins (1977; 1978). For seismic data, which are super-Gaussian in nature, he suggested to maximize the following cost:

$$\frac{\frac{1}{B} \sum_{b=1}^{B} |y_{n-b+1}|^4}{\left(\frac{1}{B} \sum_{b=1}^{B} |y_{n-b+1}|^2\right)^2} \tag{21}$$

This deconvolution scheme seeks the smallest number of large spikes consistent with the data, thus maximizing the order or, equivalently, minimizing the entropy or disorder in the data, Walden (1985). Note that the equation (21) has the statistical form of sample kurtosis and the expression is scale-invariant. Later, in the year 1979, Gray generalized the Wiggins' proposal with two degrees of freedom as follows, Gray (1979b):

$$J_{med}^{(p,q)} \equiv \frac{\frac{1}{B} \sum_{b=1}^{B} |y_{n-b+1}|^p}{\left(\frac{1}{B} \sum_{b=1}^{B} |y_{n-b+1}|^q\right)^{\frac{p}{q}}} \tag{22}$$

The criterion was rigorously investigated in Donoho (1980), where Donoho developed general rules for designing MED-type estimators. Several cases of MED, in the context of blind deconvolution of seismic data, have appeared in the literature, like $J_{med}^{(2,1)}$ in Ooe & Ulrych (1979), $J_{med}^{(4,2)}$ in Wiggins (1977), $\lim_{\bullet \to 0} J_{med}^{(p-\bullet,p)}$ in Claerbout (1977), $J_{med}^{(p,2)}$ in Gray (1978), and $J_{med}^{(2p,p)}$ in Gray (1979a).

In the derivation of the criterion (22), it is assumed that the original signal a_n, which is primary reflection coefficients in geophysical system or transmitted data in communication systems, can be modeled as realization of independent non-Gaussian process with distribution

$$p_A(a; \bullet) = \frac{\bullet}{2 \bullet \Gamma\left(\frac{1}{\bullet}\right)} \exp\left(-\frac{|a|^\bullet}{\bullet \bullet}\right) \tag{23}$$

where signal a_n is real-valued, $\bullet$ is the shape parameter, $\bullet$ is the scale parameter, and $\Gamma(\cdot)$ is the Gamma function. This family covers a wide range of distributions. The certain event

($\bullet = 0$), double exponential ($\bullet = 1$), Gaussian ($\bullet = 2$), and uniform distributions ($\bullet \to \infty$) are all members. For geophysical deconvolution problem, we have the range $0.6 \leq \bullet \leq 1.5$, and for communication system where the signals are uniformly distributed we have ($\bullet \to \infty$). Although, signals in communication are discrete, the equation (23) is still good to approximate some densely and uniformly distributed signal.

In the context of geophysics, where the primary coefficient a_n is super-Gaussian, maximizing the criterion (23) drives the distribution of deconvolved sequence y_n away from $p_y(y_n; p)$ towards $p_y(y_n; q)$, where $p > q$. However, for the communication blind equalization problem, the underlying distribution of the transmitted (possibly pulse amplitude modulated) data symbols are closer to a uniform density (sub-Gaussian) and thus we would minimize the cost (23) with $p > q$. We have the following cost for blind equalization of communication channel:

$$
w^\dagger = \begin{cases} \arg\min_w J_{\text{med}}^{(p,q)}, & \text{if } p > q, \\ \arg\max_w J_{\text{med}}^{(p,q)}, & \text{if } p < q. \end{cases} \tag{24}
$$

The feasibility of (24) for blind equalization of digital signals has been studied in Satorius & Mulligan (1992; 1993) and Benedetto et al. (2008). In Satorius & Mulligan (1992), implementing (24) with $p > q$, the following adaptive algorithm was obtained:

$$
w_{n+1} = w_n + \bullet \sum_{k=1}^{\hat{B}} \left(\frac{\sum_{b=1}^{B} |y_{n-b+1}|^p}{\sum_{b=1}^{B} |y_{n-b+1}|^q} |y_{n-k+1}|^{q-2} - |y_{n-k+1}|^{p-2} \right) y_{n-k+1}^* x_{n-k+1}, \tag{25}
$$

In the sequel, we will refer to (25) as Satorius-Mulligan algorithm (SMA). Also, for a detailed discussion on the stochastic approximate realization of MED, refer to Walden (1988).

4.3 Constant modulus criterion

The most popular and widely studied blind equalization criterion is the constant modulus criterion, Godard (1980); Treichler & Agee (1983); Treichler & Larimore (1985); it is given by

$$
J_{\text{cm}} = E\left[\left(|y_n|^2 - R_{\text{cm}} \right)^2 \right], \tag{26}
$$

where $R_{\text{cm}} = E[|a|^4] / E[|a|^2]$ is a statistical constant usually termed as dispersion constant. For an input signal that has a constant modulus $|a_n| = \sqrt{R_{\text{cm}}}$, the criterion penalizes output samples y_n that do not have the desired constant modulus characteristics. This modulus restoral concept has a particular advantage in that it allows the equalizer to be adapted independent of carrier recovery. Because the cost is insensitive to the phase of y_n, the equalizer adaptation can occur independently and simultaneously with the operation of the carrier recovery system. This property also makes it applicable to analog modulation signals with constant amplitude such as those using frequency or phase modulation, see Treichler & Larimore (1985). The stochastic gradient-descent minimization of (26) yields the following algorithm:

$$
w_{n+1} = w_n + \bullet \left(R_{\text{cm}} - |y_n|^2 \right) y_n^* x_n, \tag{27}
$$

which is usually termed as constant modulus algorithm (CMA). Note that, considering $\hat{B} = 1$, $p = 4, q = 2$ and large B, SMA (25) reduces to CMA.

If data symbols are independent and identically distributed, noise is negligible and the length of the equalizer is infinite then after some calculations, the CM cost may be expressed as a function of joint channel-equalizer impulse response coefficients as follows:

$$J_{cm} = \left(E[|a_n|^4] - 2E[|a_n|^2]^2\right) \sum_l |t_l|^4 + 2E[|a_n|^2]^2 \left(\sum_l |t_l|^2\right)^2 - 2E[|a_n|^4] \sum_l |t_l|^2 + \text{const.}$$

(28)

As in Godard (1980), the partial derivative of J_{cm} with respect to t_k can be written as

$$\frac{\bullet J_{cm}}{\bullet t_k} = 4t_k \left(E[|a_n|^4](|t_k|^2 - 1) + 2E[|a_n|^2]^2 \sum_{l \neq k} |t_l|^2\right)$$

(29)

The minimum can be found by solving $\frac{\bullet J_{cm}}{\bullet t_k} = 0$, i.e.,

$$t_k \left(E[|a_n|^4](|t_k|^2 - 1) + 2E[|a_n|^2]^2 \sum_{l \neq k} |t_l|^2\right) = 0, \ \forall \ k$$

(30)

Unfortunately, the set of equations has an infinite number of solutions; the cost J_{cm} is thus non-convex. The solutions $T_{\mathcal{M}}$, $\mathcal{M} = 1, 2, \cdots$, can be represented as follows: all elements of the set $\{t_l\}$ are equal to zero, except $\mathcal{M}$ of them and those non-zero elements have equal magnitude of $\bullet^2_{\mathcal{M}}$ defined by

$$\bullet^2_{\mathcal{M}} = \frac{E[|a_n|^4]}{E[|a_n|^4] + 2(\mathcal{M} - 1)E[|a_n|^2]^2}$$

(31)

Among these solutions, under the condition $E[|a_n|^4] < 2E[|a_n|^2]^2$, the solution T_1 is that for which the energy is the largest at the equalizer output and ISI is zero. The absolute minimum of J_{cm} is therefore reached in the case of zero IS1.

4.4 Shtrom-Fan criterion

In the year 1998, Shtrom and Fan presented a class of cost functions for achieving blind equalization which were solely the function of $\{t\}$ parameters, see Shtrom & Fan (1998). They suggested to minimize the difference between any two norms of the joint channel-equalizer impulse response, each raised to the same power, i.e.,

$$J_{sf} = \left(\sum_l |t_l|^p\right)^{r/p} - \left(\sum_l |t_l|^q\right)^{r/q}, \ p < q \text{ and } r > 0$$

(32)

where $p, q, r \in \Re$. This proposal was based on the following property of vector norms:

$$\lim_{s \to 0} \sqrt[s]{\sum_l |t_l|^s} \geq \cdots \geq \sqrt[p]{\sum_l |t_l|^p} \geq \sqrt[q]{\sum_l |t_l|^q} \geq \cdots \geq \lim_{m \to \infty} \sqrt[m]{\sum_l |t_l|^m}$$

(33)

where $p < q$ and equality occurs if and only if $t_l = \pm \cdot_{l-k}$, $k \in \mathbb{Z}$, which is precisely the zero-forcing condition. From the above there is a multitude of cost functions to choose from. From (32), we have following possibilities to minimize:

$$\sum_l |t_l| - \max_l\{|t_l|\}, \quad (p = 1, q \to \infty, r = 1) \tag{34a}$$

$$\left(\sum_l |t_l|\right)^2 - \sum_l |t_l|^2, \quad (p = 1, q = r = 2) \tag{34b}$$

$$\left(\sum_l |t_l|^2\right)^2 - \sum_l |t_l|^4, \quad (p = 2, q = r = 4) \tag{34c}$$

$$\left(\sum_l |t_l|^2\right)^3 - \sum_l |t_l|^6, \quad (p = 2, q = r = 6) \tag{34d}$$

$$\left(\sum_l |t_l|^4\right)^2 - \sum_l |t_l|^8, \quad (p = 4, q = r = 8). \tag{34e}$$

Some of these cost functions are easily implementable, whereas others are not.

Consider $p = 2$ and $q = r = 2m$ in (32) to obtain a subclass:

$$J_{\text{sf}}^{\text{sub}} = \left(\sum_l |t_l|^2\right)^m - \sum_l |t_l|^{2m} \tag{35}$$

This subclass is not convex, although it is potentially unimodal in t domain and easily implementable. As in Shtrom & Fan (1998), the partial derivative of $J_{\text{sf}}^{\text{sub}}$ with respect to t_k can be written as

$$\frac{\bullet\, J_{\text{sf}}^{\text{sub}}}{\bullet\, t_k} = 2mt_k\left(\left(\sum_l |t_l|^2\right)^{m-1} - |t_k|^{2(m-1)}\right) \tag{36}$$

The equation (36) has two solutions, one of which corresponds to $t_l = 0$, $\forall l$. This solution will not occur if a constraint is imposed. The other solution is the minimum corresponding to zero-forcing condition. This is seen from (36) as $\sum_l |t_l|^2 = |t_k|^2$, which can only hold when t has at most one nonzero element, i.e., the desired delta function. Now compare this result with that of constant modulus in equation (31) which contains multiple nonzero-forcing solutions. It means, in contrast to CMA, it is less likely to have local minima in $J_{\text{sf}}^{\text{sub}}$ in equalizer domain.

The cost functions (34a), in their current form, are not directly applicable in real scenario as we have no information of $\{t\}$'s. These costs need to be converted from functions of $\{t\}$'s to functions of y_n's. As in Shtrom & Fan (1998), we can show that

$$\sum_l |t_l|^2 = \frac{C_{1,1}^{y_n}}{C_{1,1}^a} = \frac{E[|y_n|^2]}{E[|a|^2]} \tag{37a}$$

$$\sum_l |t_l|^4 = \frac{C_{2,2}^{y_n}}{C_{2,2}^a} = \frac{E[|y_n|^4] - 2\,E[|y_n|^2]^2}{E[|a|^4] - 2\,E[|a|^2]^2} \tag{37b}$$

$$\sum_l |t_l|^6 = \frac{C_{3,3}^{y_n}}{C_{3,3}^a} = \frac{E[|y_n|^6] - 9\,E[|y_n|^4]E[|y_n|^2] + 12\,E[|y_n|^2]^3}{E[|a|^6] - 9\,E[|a|^4]E[|a|^2] + 12\,E[|a|^2]^3} \tag{37c}$$

where $C_{p,q}^z$ is $(p+q)$th order cumulant of complex random variable defined as follows:

$$C_{p,q}^z = \text{cumulant}\left(\underbrace{z, \cdots, z}_{p\ \text{terms}};\ \underbrace{z^*, \cdots, z^*}_{q\ \text{terms}}\right) \tag{38}$$

Using (37a) and assuming $m = 2$, we obtain the following expression for J_{sf}^{sub}:

$$J_{sf}^{sub} = \left(\frac{E[|y_n|^2]}{E[|a|^2]} \right)^2 - \frac{E[|y_n|^4] - 2E[|y_n|^2]^2}{E[|a|^4] - 2E[|a|^2]^2} \tag{39}$$

Minimizing (39) with respect to coefficients w^*, we obtain the following adaptive algorithm:

$$w_{n+1} = w_n + \cdot \left(\frac{R_{cm}}{P_a} \widehat{E}[|y_n|^2] - |y_n|^2 \right) y_n^* x_n, \tag{40a}$$

$$\widehat{E}[|y_{n+1}|^2] = \widehat{E}[|y_n|^2] + \frac{1}{n} \left(|y_n|^2 - \widehat{E}[|y_n|^2] \right) \tag{40b}$$

where P_a is the average energy of signal a_n and R_{cm} is the same statistical constant as we defined in CMA. Note that the algorithm requires an iterative estimate of equalizer output energy. We will refer to (40) as Shtrom-Fan Algorithm (SFA).

4.5 Shalvi-Weinstein criterion

Earlier to Shtrom and Fan, in the year 1990, Shalvi and Weinstein suggested a criterion that laid the theoretical foundation to the problem of blind equalization, see Shalvi & Weinstein (1990). They demonstrated that the condition of equality between the PDF's of the transmitted and equalized signals, due to BGR theorem Benveniste et al. (1980a;b), was excessively tight. Under the similar assumptions, as laid by Benveniste *et al.*, they demonstrated that it is possible to perform blind equalization by satisfying the condition $E[|y_n|^2] = E[|a_n|^2]$ and ensuring that a nonzero cumulant of order higher than 2 of a_n and y_n are equal.

For a two dimensional signal a_n with four-quadrant symmetry (i.e., $E[a_n^2] = 0$), they suggested to maximize the following unconstrained cost function (which involved second and fourth order cumulants):

$$J_{sw} = sgn\left[C_{2,2}^{a_n}\right] \left(C_{2,2}^{y_n} + (\cdot_1 + 2) \left(C_{1,1}^{y_n} \right)^2 + 2 \cdot_2 C_{1,1}^{y_n} \right) \tag{41a}$$

$$= sgn\left[C_{2,2}^{a_n}\right] \left(E\left[|y_n|^4\right] + \cdot_1 E\left[|y_n|^2\right]^2 + 2 \cdot_2 E\left[|y_n|^2\right] \right) \tag{41b}$$

where $\cdot_1$ and $\cdot_2$ are some statistical constants. The corresponding stochastic gradient algorithm is given by

$$w_{n+1} = w_n + \cdot sgn[C_{2,2}^{a_n}] \left(\cdot_1 \widehat{E}[|y_n|^2] + |y_n|^2 + \cdot_2 \right) y_n^* x_n, \tag{42a}$$

$$\widehat{E}[|y_{n+1}|^2] = \widehat{E}[|y_n|^2] + \frac{1}{n} \left(|y_n|^2 - \widehat{E}[|y_n|^2] \right) \tag{42b}$$

where $(sgn[C_{2,2}^{a_n}] = -1)$ due to the sub-Gaussian nature of digital signals. The above algorithm is usually termed as Shalvi-Weinstein algorithm (SWA). Note that SWA unifies CMA and SFA (i.e., the specific case in Equation (40)). Substituting $\cdot_1 = 0$ and $\cdot_2 = -R_{cm}$ in SWA, we obtain CMA (27). Similarly, substituting $\cdot_2 = 0$ and $\cdot_1 = -R_{cm}/P_a$ in SWA, we obtain SFA (40). Note that the Shtrom-Fan criterion appears to be the generalization of Shalvi-Weinstein criterion with cumulants of generic orders.

5. Blind equalization of APSK signal: Employing MED principle

5.1 Designing a blind cost function

We employ MED principle and use the PDFs of transmitted amplitude-phase shift-keying (APSK) and ISI-affected received signal to design a cost function for blind equalization. Consider a continuous APSK signal, where signal alphabets $\{a_R + \bullet a_I\} \in \mathcal{A}$ are assumed to be uniformly distributed over a circular region of radius R_a and center at the origin. The joint PDF of a_R and a_I is given by (refer to Fig. 3(a))

$$p_{\mathcal{A}}(a_R + \bullet a_I) = \begin{cases} \dfrac{1}{\bullet R_a^2}, & \sqrt{a_R^2 + a_I^2} \leq R_a, \\ 0, & \text{otherwise.} \end{cases} \tag{43}$$

Now consider the transformation $\mathcal{Y} = \sqrt{a_R^2 + a_I^2}$ and $\Theta = \bullet (a_R, a_I)$, where $\mathcal{Y}$ is the modulus and $\bullet ()$ denotes the angle in the range $(0, 2\bullet)$ that is defined by the point (i, j). The joint distribution of the modulus $\mathcal{Y}$ and Θ can be obtained as $p_{\mathcal{Y},\Theta}(\tilde{y}, \tilde{\bullet}) = \tilde{y}/(\bullet R_a^2)$, $\tilde{y} \geq 0, 0 \leq \tilde{\bullet} < 2\bullet $. Since $\mathcal{Y}$ and Θ are independent, we obtain a triangular distribution for $\mathcal{Y}$ given by $p_{\mathcal{Y}}(\tilde{y} : H_0) = 2\tilde{y}/R_a^2$, $\tilde{y} \geq 0$, where H_0 denotes the hypothesis that signal is distortion-free. Let $\mathcal{Y}_n, \mathcal{Y}_{n-1}, \cdots, \mathcal{Y}_{n-N+1}$ be a sequence, of size N, obtained by taking modulus of randomly

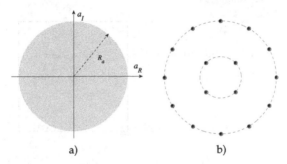

Fig. 2. a) A continuous APSK, and b) a discrete practical 16APSK.

generated distortion-free signal alphabets $\mathcal{A}$, where subscript n indicates discrete time index. Let $\mathcal{Z}_1, \mathcal{Z}_2, \cdots, \mathcal{Z}_N$ be the order statistic of sequence $\{\mathcal{Y}\}$. Let $p_{\mathcal{Y}}(\tilde{y}_n, \tilde{y}_{n-1}, \cdots, \tilde{y}_{n-N+1} : H_0)$ be an N-variate density of the continuous type, then, under the hypothesis H_0, we obtain

$$p_{\mathcal{Y}}(\tilde{y}_n, \tilde{y}_{n-1}, \cdots, \tilde{y}_{n-N+1} : H_0) = \frac{2^N}{R_a^{2N}} \prod_{k=1}^{N} \tilde{y}_{n-k+1}. \tag{44}$$

Next we find $p_{\mathcal{Y}}^*(\tilde{y}_n, \tilde{y}_{n-1}, \cdots, \tilde{y}_{n-N+1} : H_0)$ as follows:

$$p_{\mathcal{Y}}^*(\tilde{y}_n, \tilde{y}_{n-1}, \cdots, \tilde{y}_{n-N+1} : H_0) = \int_0^{\infty} p_{\mathcal{Y}}(\bullet \tilde{y}_n, \bullet \tilde{y}_{n-1}, \cdots, \bullet \tilde{y}_{n-N+1} : H_0) \bullet^{N-1} \mathrm{d}\bullet$$

$$= \frac{2^N}{R_a^{2N}} \prod_{k=1}^{N} \tilde{y}_{n-k+1} \int_0^{R_a/\tilde{z}_N} \bullet^{2N-1} \mathrm{d}\bullet = \frac{2^{N-1}}{N (\tilde{z}_N)^{2N}} \prod_{k=1}^{N} \tilde{y}_{n-k+1}, \tag{45}$$

where $\tilde{z}_1, \tilde{z}_2, \cdots, \tilde{z}_N$ are the order statistic of elements $\tilde{y}_n, \tilde{y}_{n-1}, \cdots, \tilde{y}_{n-N+1}$, so that $z_1 = \min\{\tilde{y}\}$ and $z_N = \max\{\tilde{y}\}$. Now consider the next hypothesis (H_1) that signal suffers with multi-path interference as well as with additive Gaussian noise (refer to Fig. 3(b)). Due to which, the in-phase and quadrature components of the received signal are modeled as normal distributed; owing to central limit theorem, it is theoretically justified. It means that the modulus of the received signal follows Rayleigh distribution,

$$p_y(\tilde{y} : H_1) = \frac{\tilde{y}}{\bullet_{\tilde{y}}^2} \exp\left(-\frac{\tilde{y}^2}{2\bullet_{\tilde{y}}^2}\right), \tilde{y} \geq 0, \bullet_{\tilde{y}} > 0. \tag{46}$$

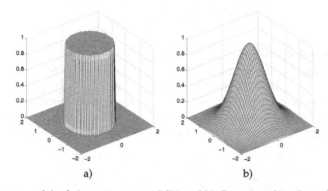

a)   b)

Fig. 3. PDFs (not to scale) of a) continuous APSK and b) Gaussian distributed received signal.

The N-variate densities $p_y(\tilde{y}_n, \tilde{y}_{n-1}, \cdots, \tilde{y}_{n-N+1} : H_1)$ and $p_{\tilde{y}}^*(\tilde{y}_n, \tilde{y}_{n-1}, \cdots, \tilde{y}_{n-N+1} : H_1)$ are obtained as

$$p_y(\tilde{y}_n, \tilde{y}_{n-1}, \cdots, \tilde{y}_{n-N+1} : H_1) = \frac{1}{\bullet_{\tilde{y}}^{2N}} \prod_{k=1}^N \tilde{y}_{n-k+1} \exp\left(-\frac{\tilde{y}_{n-k+1}^2}{2\bullet_{\tilde{y}}^2}\right) \tag{47}$$

$$p_{\tilde{y}}^*(\tilde{y}_n, \tilde{y}_{n-1}, \cdots, \tilde{y}_{n-N+1} : H_1) = \frac{\prod_{k=1}^N \tilde{y}_{n-k+1}}{\bullet_{\tilde{y}}^{2N}} \int_0^\infty \exp\left(-\frac{\bullet^2 \sum_{k'=1}^N \tilde{y}_{n-k'+1}^2}{2\bullet_{\tilde{y}}^2}\right) \bullet^{2N-1} \mathrm{d}\bullet \tag{48}$$

Substituting $u = \frac{1}{2} \bullet^2 \bullet_{\tilde{y}}^{-2} \sum_{k'=1}^N \tilde{y}_{n-k'+1}^2$, we obtain

$$p_{\tilde{y}}^*(\tilde{y}_n, \tilde{y}_{n-1}, \cdots, \tilde{y}_{n-N+1} : H_1) = \frac{2^{N-1}\Gamma(N)}{\left(\sum_{k=1}^N \tilde{y}_{n-k+1}^2\right)^N} \prod_{k=1}^N \tilde{y}_{n-k+1} \tag{49}$$

The scale-invariant uniformly most powerful test of $p_{\tilde{y}}^*(\tilde{y}_n, \tilde{y}_{n-1}, \cdots, \tilde{y}_{n-N+1} : H_0)$ against $p_{\tilde{y}}^*(\tilde{y}_n, \tilde{y}_{n-1}, \cdots, \tilde{y}_{n-N+1} : H_1)$ provides us, see Sidak et al. (1999):

$$O(\tilde{y}_n) = \frac{p_{\tilde{y}}^*(\tilde{y}_n, \tilde{y}_{n-1}, \cdots, \tilde{y}_{n-N+1} : H_0)}{p_{\tilde{y}}^*(\tilde{y}_n, \tilde{y}_{n-1}, \cdots, \tilde{y}_{n-N+1} : H_1)} = \frac{1}{N!} \left[\frac{\sum_{k=1}^N \tilde{y}_{n-k+1}^2}{\tilde{z}_N^2}\right]^N \begin{matrix} H_0 \\ \bullet \\ H_1 \end{matrix} C \tag{50}$$

where C is a threshold. Assuming large N, we can approximate $\frac{1}{N}\sum_{k=1}^{N}\tilde{y}_{n-k+1}^2 \approx \mathsf{E}\left[|y_n|^2\right]$. It helps obtaining a statistical cost for the blind equalization of APSK signal as follows:

$$w^{\dagger} = \arg\max_{w} \frac{\mathsf{E}\left[|y_n|^2\right]}{\left(\max\{|y_n|\}\right)^2} \tag{51}$$

Based on the previous discussion, maximizing cost (51) can be interpreted as determining the equalizer coefficients, w, which drives the distribution of its output, y_n, away from Gaussian distribution toward uniform, thus removing successfully the interference from the received APSK signal. Note that the above result (51) may be obtained directly from (24) by substituting $p = 2$ and $q \rightarrow \infty$, see Abrar & Nandi (2010b).

5.2 Admissibility of the proposed cost

The cost (51) demands maximizing equalizer output energy while minimizing the largest modulus. Since the largest modulus of transmitted signal a_n is R_a, incorporating this *a priori* knowledge, the unconstrained cost (51) can be written in a constrained form as follows:

$$w^{\dagger} = \arg\max_{w} \mathsf{E}\left[|y_n|^2\right] \quad \text{s.t.} \quad \max\{|y_n|\} \leq R_a. \tag{52}$$

By incorporating R_a, it would be possible to recover the true energy of signal a_n upon successful convergence. Also note that $\max\{|y_n|\} = R_a \sum_l |t_l|$ and $\mathsf{E}\left[|y_n|^2\right] = P_a \sum_l |t_l|^2$. Based on which, we note that the cost (52) is quadratic, and the feasible region (constraint) is a convex set (proof of which is similar to that in Equation (9)). The problem, however, is non-convex and may have multiple local maxima. Nevertheless, we have the following theorem:

Theorem: Assume $w^{\dagger}$ is a local optimum in (52), and $t^{\dagger}$ is the corresponding total channel equalizer impulse-response and channel noise is negligible. Then it holds $|t_l| = \bullet_{l-l^{\dagger}}$.

Proof: Without loss of generality we assume that the channel and equalizer are real-valued. We re-write (52) as follows:

$$w^{\dagger} = \arg\max_{w} \sum_l t_l^2 \quad \text{s.t.} \quad \sum_l |t_l| \leq 1. \tag{53}$$

Now consider the following quadratic problem in t domain

$$t^{\dagger} = \arg\max_{t} \sum_l t_l^2 \quad \text{s.t.} \quad \sum_l |t_l| \leq 1. \tag{54}$$

Assume $t^{(f)}$ is a feasible solution to (54). We have

$$\sum_l t_l^2 \leq \left(\sum_l |t_l|\right)^2 \leq 1 \tag{55}$$

and

$$\left(\sum_l |t_l|\right)^2 = \sum_l t_l^2 + \sum_{l_1}\sum_{l_2,\, l_2 \neq l_1} |t_{l_1} t_{l_2}| \tag{56}$$

The first equality in (55) is achieved if and only if all cross terms in (56) are zeros. Now assume that $t^{(k)}$ is a local optimum of (54), i.e., the following proposition holds

$$\exists \bullet > 0, \ \forall t^{(f)}, \ \|t^{(f)} - t^{(k)}\|_2 \leq \bullet \tag{57}$$

$\Rightarrow \sum_l (t_l^{(k)})^2 \geq \sum_l (t_l^{(f)})^2$. Suppose $t^{(k)}$ does not satisfy the **Theorem**. Consider $t^{(c)}$ defined by

$$t_{l_1}^{(c)} = t_{l_1}^{(k)} + \frac{\bullet}{\sqrt{2}},$$

$$t_{l_2}^{(c)} = t_{l_2}^{(k)} - \frac{\bullet}{\sqrt{2}},$$

and $t_l^{(c)} = t_l^{(k)}$, $l \neq l_1, l_2$. We also assume that $t_{l_2}^{(k)} < t_{l_1}^{(k)}$. Next, we have $\|t^{(c)} - t^{(k)}\|_2 = \bullet$, and $\sum_l |t_l^{(c)}| = \sum_l |t_l^{(k)}| \leq 1$. However, one can observe that

$$\sum_l (t_l^{(k)})^2 - \sum_l (t_l^{(c)})^2 = \sqrt{2} \bullet \left(t_{l_2}^{(k)} - t_{l_1}^{(k)} \right) - \bullet^2 < 0, \tag{58}$$

which means $t^{(k)}$ is not a local optimum to (54). Therefore, we have shown by a counterexample that all local maxima of (54) should satisfy the **Theorem**.

5.3 Adaptive optimization of the proposed cost

For a stochastic gradient-based adaptive implementation of (52), we need to modify it to involve a *differentiable* constraint; one of the possibilities is

$$w^\dagger = \arg\max_w \mathrm{E}\left[|y_n|^2 \right] \ \text{s.t.} \ \mathrm{fmax}(R_a, |y_n|) = R_a, \tag{59}$$

where we have used the following identity (below $a, b \in \mathbb{C}$):

$$\mathrm{fmax}(|a|, |b|) \equiv \frac{\big||a| + |b|\big| + \big||a| - |b|\big|}{2} = \begin{cases} |a|, & \text{if } |a| \geq |b| \\ |b|, & \text{otherwise.} \end{cases} \tag{60}$$

The function fmax is differentiable, viz

$$\frac{\bullet \, \mathrm{fmax}(|a|, |b|)}{\bullet \, a^*} = \frac{a\big(1 + \mathrm{sgn}(|a| - |b|)\big)}{4|a|} = \begin{cases} a/(2|a|), & \text{if } |a| > |b| \\ 0, & \text{if } |a| < |b| \end{cases} \tag{61}$$

If $|y_n| < R_a$, then the cost (59) simply maximizes output energy. However, if $|y_n| > R_a$, then the constraint is violated and the new update w_{n+1} is required to be computed such that the magnitude of *a posteriori* output $w_{n+1}^H x_n$ becomes smaller than or equal to R_a. Next, employing Lagrangian multiplier, we get

$$w^\dagger = \arg\max_w \left\{ \mathrm{E}[|y_n|^2] + \bullet \, (\mathrm{fmax}\,(R_a, |y_n|) - R_a) \right\}. \tag{62}$$

The stochastic approximate gradient-based optimization of $w^\dagger = \arg\max_w \mathrm{E}[\mathcal{J}]$ is realized as $w_{n+1} = w_n + \bullet \bullet \mathcal{J}/\bullet w_n^*$, where $\bullet > 0$ is a small step-size. Differentiating (62) with respect to w_n^* gives

$$\frac{•\,|y_n|^2}{•\,w_n^*} = \frac{•\,|y_n|^2}{•\,y_n}\frac{•\,y_n}{•\,w_n^*} = y_n^* x_n$$

and

$$\frac{•\,\text{fmax}(R_a, |y_n|)}{•\,w_n^*} = \frac{•\,\text{fmax}(R_a, |y_n|)}{•\,y_n}\frac{•\,y_n}{•\,w_n^*} = \frac{g_n y_n^*}{4|y_n|} x_n,$$

where $g_n \equiv 1 + \text{sgn}(|y_n| - R_a)$; we obtain $w_{n+1} = w_n + • \left(1 + • \, g_n/(4|y_n|)\right) y_n^* x_n$. If $|y_n| < R_a$, then $g_n = 0$ and $w_{n+1} = w_n + • \, y_n^* x_n$. Otherwise, if $|y_n| > R_a$, then $g_n = 2$ and

$$w_{n+1} = w_n + • \left(1 + • /(2\,|y_n|)\right) y_n^* x_n.$$

As mentioned earlier, in this case, we have to compute $•$ such that $w_{n+1}^H x_n$ lies inside the circular region without sacrificing the output energy. Such an update can be realized by minimizing $|y_n|^2$ while satisfying the Bussgang condition, see Bellini (1994). Note that the satisfaction of Bussgang condition ensures recovery of the true signal energy upon successful convergence. One of the possibilities is $• = -2(1 + •)|y_n|$, $• > 0$, which leads to

$$w_{n+1} = w_n + • \, (-•) y_n^* x_n.$$

The Bussgang condition requires

$$\underbrace{E\left[y_n y_{n-i}^*\right]}_{|y_n|<R_a} + \underbrace{(-•)\, E\left[y_n y_{n-i}^*\right]}_{|y_n|>R_a} = 0, \ \forall i \in \mathbb{Z} \tag{63}$$

In steady-state, we assume $y_n = a_{n-•} + u_n$, where u_n is convolutional noise. For $i \neq 0$, (63) is satisfied due to uncorrelated a_n and independent and identically distributed samples of u_n. Let a_n comprises M distinct symbols on L moduli $\{R_1, \cdots, R_L\}$, where $R_L = R_a$ is the largest modulus. Let M_i denote the number of unique (distortion-free) symbols on the ith modulus, i.e., $\sum_{l=1}^{L} M_l = M$. With negligible u_n, we solve (63) for $i = 0$ to get

$$M_1 R_1^2 + \cdots + M_{L-1} R_{L-1}^2 + \frac{1}{2} M_L R_L^2 - \frac{•}{2} M_L R_L^2 = 0 \tag{64}$$

The last two terms indicate that, when $|y_n|$ is close to R_L, it would be equally likely to update in either direction. Noting that $\sum_{l=1}^{L} M_l R_l^2 = M P_a$, the simplification of (64) gives

$$• = 2\frac{M}{M_L}\frac{P_a}{R_a^2} - 1. \tag{65}$$

The use of (65) ensures recovery of true signal energy upon successful convergence. Finally the proposed algorithm is expressed as

$$w_{n+1} = w_n + • \, \text{f}(y_n)\, y_n^* x_n,$$
$$\text{f}(y_n) = \begin{cases} 1, & \text{if } |y_n| \le R_a \\ -•, & \text{if } |y_n| > R_a. \end{cases} \tag{66}$$

Note that the error-function $f(y_n)y_n^*$ has 1) finite derivative at the origin, 2) is increasing for $|y_n| < R_a$, 3) decreasing for $|y_n| > R_a$ and 4) insensitive to phase/frequency offset errors. In Baykal et al. (1999), these properties 1)-4) have been regarded as essential features of a constant modulus algorithm; this motivates us to denote (66) as $\bullet$CMA.

5.4 Stability of the derived adaptive algorithm

In this Section, we carry out a deterministic stability analysis of $\bullet$CMA for any bounded magnitude received signal. The analysis relies on the analytical framework of Rupp & Sayed (2000). We shall assume that the successive regression vectors $\{x_i\}$ are nonzero and also uniformly bounded from above and from below. We update the equalizer only when its output amplitude is higher than certain threshold; we stop the update otherwise.

In our analysis, we assume that the threshold is R_a. So we only consider those updates when $|y_n| > R_a$; we extract and denote the active update steps with time index k. We study the following form:

$$w_{k+1} = w_k + \bullet_k \Phi_k^* x_k, \quad \Phi_k \neq 0, \; k = 0, 1, 2, \cdots \tag{67}$$

where $\Phi_k \equiv \Phi(y_k) = f(y_k)y_k$. Let w_* denote vector of the optimal equalizer and let $z_k = w_*^H x_k = a_{k-\bullet}$ is the optimal output so that $|z_k| = R_a$. Define the *a priori* and *a posteriori* estimation errors

$$e_k^a := z_k - y_k = \overline{w}_k^H x_k$$

$$e_k^p := z_k - s_k = \overline{w}_{k+1}^H x_k \tag{68}$$

where $\overline{w}_k = w_* - w_k$. We assume that $|e_k^a|$ is small and equalizer is operating in the vicinity of w_*. We introduce a function $\bullet(x, y)$:

$$\bullet(x, y) := \frac{\Phi(y) - \Phi(x)}{x - y} = \frac{f(y)\, y - f(x)\, x}{x - y}, \quad (x \neq y) \tag{69}$$

Using $\bullet(x, y)$ and simple algebra, we obtain

$$\Phi_k = f(y_k)\, y_k = \bullet(z_k, y_k) e_k^a \tag{70}$$

$$e_k^p = \left(1 - \frac{\bullet_k}{\bullet_k} \bullet(z_k, y_k)\right) e_k^a \tag{71}$$

where $\overline{\bullet}_k = 1/\|x_k\|^2$. For the stability of adaptation, we require $|e_k^p| < |e_k^a|$. To ensure it, we require to guarantee the following for all possible combinations of z_k and y_k:

$$\left|1 - \frac{\bullet_k}{\bullet_k} \bullet(z_k, y_k)\right| < 1, \; \forall k \tag{72}$$

Now we require to prove that the real part of the function $\bullet(z_k, y_k)$ defined by (69) is positive and bounded from below. Recall that $|z_k| = R_a$ and $|y_k| > R_a$. We start by writing $z_k/y_k = re^{j\bullet}$ for some $r < 1$ and for some $\bullet \in [0, 2\bullet)$. Then expression (69) leads to

$$\bullet(z_k, y_k) = \frac{\overbrace{f(y_k)}^{(=-\bullet)}\, y_k - \overbrace{f(z_k)}^{(=0)}\, z_k}{z_k - y_k} = \frac{\bullet\, y_k}{y_k - z_k} = \frac{\bullet}{1 - re^{j\bullet}}. \tag{73}$$

It is important for our purpose to verify whether the real part of $\bullet\,/(1-re^{j\bullet})$ is positive. For any fixed value of r, we allow the angle $\bullet$ to vary from zero to $2\bullet$, then the term $\bullet\,/(1-re^{j\bullet})$ describes a circle in the complex plane whose least positive value is $\bullet\,/(1+r)$, obtained for $\bullet=\bullet$, and whose most positive value is $\bullet\,/(1-r)$, obtained for $\bullet=0$. This shows that for $r\in(0,1)$, the real part of the function $\bullet(z_k,y_k)$ lies in the interval

$$\frac{\bullet}{1+r}\le \bullet_R(z_k,y_k)\le\frac{\bullet}{1-r} \tag{74}$$

Referring to Fig. 4, note that the function $\bullet(z_k,y_k)$ assumes values that lie inside a circle in the

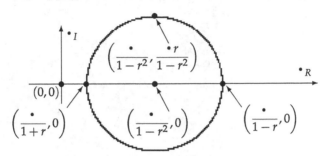

Fig. 4. Plot of $\bullet(z_k,y_k)$ for arbitrary $\bullet$, r and $\bullet\in[0,2\bullet)$.

right-half plane. From this figure, we can obtain the following bound for the imaginary part of $\bullet(z_k,y_k)$ (that is $\bullet_I(z_k,y_k)$):

$$-\frac{\bullet\,r}{1-r^2}\le \bullet_I(z_k,y_k)\le\frac{\bullet\,r}{1-r^2}. \tag{75}$$

Let A and B be any two positive numbers satisfying

$$A^2+B^2<1. \tag{76}$$

We need to find a $\bullet_k$ that satisfies

$$\left|\frac{\bullet_k}{\bullet_k}\bullet_I(z_k,y_k)\right|<A \ \Rightarrow\ \bullet_k<\frac{A\,\bar{\bullet}_k}{|\bullet_I(z_k,y_k)|} \tag{77}$$

and

$$\left|1-\frac{\bullet_k}{\bullet_k}\bullet_R(z_k,y_k)\right|<B \ \Rightarrow\ \bullet_k>\frac{(1-B)\bar{\bullet}_k}{\bullet_R(z_k,y_k)} \tag{78}$$

Combining (77) and (78), we obtain

$$0<\frac{(1-B)\bar{\bullet}_k}{\bullet_R(z_k,y_k)}<\bullet_k<\frac{A\,\bar{\bullet}_k}{|\bullet_I(z_k,y_k)|}<1 \tag{79}$$

Using the extremum values of $\bullet_R(z_k,y_k)$ and $\bullet_I(z_k,y_k)$, we obtain

$$\frac{(1+r)(1-B)}{\bullet\,\|x_k\|^2}<\bullet_k<\frac{(1-r^2)A}{\bullet\,r\|x_k\|^2} \tag{80}$$

We need to guarantee that the upper bound in the above expression is larger than the lower bound. This can be achieved by choosing $\{A, B\}$ properly such that

$$0 < (1 - B) < \frac{1 - r}{r} A < 1 \tag{81}$$

From our initial assumptions that the equalizer is in the vicinity of open-eye solution and $|y_k| > R_a$, we know that $r < 1$. It implies that we require to determine the *smallest* value of r which satisfies (81), or in other words, we have to determine the bound for step-size with the *largest* equalizer output amplitude. In Fig. 5(a), we plot the function $f_r := (1 - r)/r$ versus r; note that for $0.5 \le r < 1$ we have $f_r \le 1$.

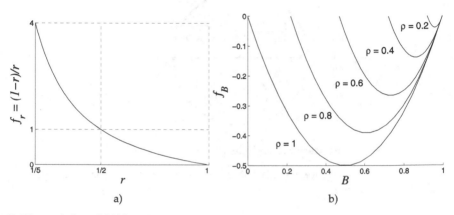

Fig. 5. Plots: a) f_r and b) f_B.

Let $\{A_o, B_o\}$ be such that $(1 - B_o) = \bullet\, A_o$, where $0 < \bullet < 1$. To satisfy (81), we need $0.5 \le r < 1$ and $\bullet < (1 - r)/r$. From (76), B_o must be such that

$$(1 - B_o)^2 \,\bullet^{-2} + B_o^2 < 1 \tag{82}$$

which reduces to the following quadratic inequality in B_o:

$$\left(1 + \bullet^{-2}\right) B_o^2 - 2\bullet^{-2} B_o + \left(\bullet^{-2} - 1\right) < 0. \tag{83}$$

If we find a B_o that satisfies this inequality, then a pair $\{A_o, B_o\}$ satisfying (76) and (81) exists. So consider the quadratic function $f_B := \left(1 + \bullet^{-2}\right) B^2 - 2\bullet^{-2} B + (\bullet^{-2} - 1)$. It has a negative minimum and it crosses the real axis at the positive roots $B^{(1)} = \left(1 - \bullet^2\right) / \left(1 + \bullet^2\right)$, and $B^{(2)} = 1$. This means that there exist many values of B, between the roots, at which the quadratic function in B evaluates to negative values (refer to Fig. 5(b)).

Hence, B_o falls in the interval $(1 - \bullet^2)/(1 + \bullet^2) < B_o < 1$; it further gives $A_o = 2\bullet / \left(1 + \bullet^2\right)$. Using $\{A_o, B_o\}$, we obtain

$$\frac{3\bullet^2}{\bullet \left(1 + \bullet^2\right) \|x_k\|^2} < \bullet_k < \frac{3\bullet}{\bullet \left(1 + \bullet^2\right) \|x_k\|^2} \tag{84}$$

Note that, $\arg\min. \bullet^2/(1+\bullet^2) = 0$, and $\arg\max. \bullet/(1+\bullet^2) = 1$. So making $\bullet = 0$ and $\bullet = 1$ in the lower and upper bounds, respectively, and replacing $\|x_k\|^2$ with $\mathsf{E}[\|x_k\|^2]$, we find the widest stochastic stability bound on $\bullet_k$ as follows:

$$0 < \bullet < \frac{3}{2\bullet \, \mathsf{E}[\|x_k\|^2]}. \tag{85}$$

The significance of (85) is that it can easily be measured from the equalizer input samples. In adaptive filter theory, it is convenient to replace $\mathsf{E}[\|x_k\|^2]$ with $\mathrm{tr}(R)$, where $R = \mathsf{E}[x_k x_k^H]$ is the autocorrelation matrix of channel observation. Also note that, when noise is negligible and channel coefficients are normalized, the quantity $\mathrm{tr}(R)$ can be expressed as the product of equalizer length (N) and transmitted signal average energy (P_a); it gives

$$0 < \bullet \bullet_{\mathrm{CMA}} < \frac{3}{2\bullet \, NP_a} \tag{86}$$

Note that the bound (86) is remarkably similar to the stability bound of complex-valued LMS algorithm (refer to expression (20)). Comparing (86) and (20), we obtain a simple and elegant relation between the step-sizes of $\bullet$ CMA and complex-valued LMS:

$$\frac{\bullet \bullet_{\mathrm{CMA}}}{\bullet_{\mathrm{LMS}}} < \frac{3}{\bullet}. \tag{87}$$

6. Simulation results

We compare $\bullet$ CMA with CMA (Equation (27)) and SFA (Equation (40)). We consider transmission of amplitude-phase shift-keying (APSK) signals over a complex voice-band channel (channel-1), taken from Picchi & Prati (1987), and evaluate the average ISI traces at SNR = 30dB. We employ a seven-tap equalizer with central spike initialization and use 8- and 16-APSK signalling. Step-sizes have been selected such that all algorithms reached steady-state requiring same number of iterations. The parameter $\bullet$ is obtained as:

$$\bullet = \begin{cases} 1.535, & \text{for 8.APSK} \\ 1.559, & \text{for 16.APSK} \end{cases} \tag{88}$$

Results are summarized in Fig. 6; note that the $\bullet$ CMA performed better than CMA and SFA by yielding much lower ISI floor. Also note that SFA performed slightly better than CMA.

Next we validate the stability bound (86). Here we consider a second complex channel (as channel-2) taken from Kennedy & Ding (1992). In all cases, the simulations were performed with $N_{\mathrm{it}} = 10^4$ iterations, $N_{\mathrm{run}} = 100$ runs, and no noise. In Fig. 7, we plot the probability of divergence (P_{div}) for three different equalizer lengths, against the normalized step-size, $\bullet_{\mathrm{norm}} = \bullet/\bullet_{\mathrm{bound}}$. The P_{div} is estimated as $P_{\mathrm{div}} = N_{\mathrm{div}}/N_{\mathrm{run}}$, where N_{div} indicates the number of times equalizer diverged. Equalizers were initialized close to zero-forcing solution. It can be seen that the bound does guarantee a stable performance when $\bullet < \bullet_{\mathrm{bound}}$.

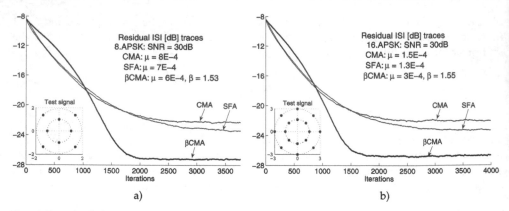

Fig. 6. Residual ISI: a) 8-APSK and b) 16-APSK. The inner and outer moduli of 8-APSK are 1.000 and 1.932, respectively. And the inner and outer moduli of 16-APSK are 1.586 and 3.000, respectively. The energies of 8-APSK and 16-APSK are 2.366 and 5.757, respectively.

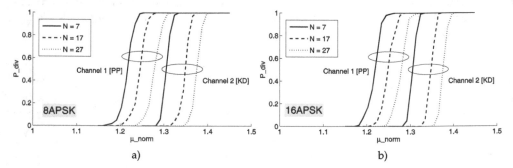

Fig. 7. Probability of divergence on channel-1 and channel-2 with three equalizer lengths, no noise, $N_{it} = 10^4$ iterations and $N_{run} = 100$ runs for a) 8-APSK and b) 16-APSK.

7. Concluding remarks

In this Chapter, we have introduced the basic concept of adaptive blind equalization in the context of single-input single-output communication systems. The key challenge of adaptive blind equalizers lies in the design of special cost functions whose minimization or maximization result in the removal of inter-symbol interference. We have briefly discussed popular criteria of equalization like Lucky, the mean square error, the minimum entropy, the constant modulus, Shalvi-Weinstein and Shtrom-Fan. Most importantly, based on minimum entropy deconvolution principle, the idea of designing specific cost function for the blind equalization of given transmitted signal is described in detail. We have presented a case study of amplitude-phase shift-keying signal for which a cost function is derived and corresponding adaptive algorithm is obtained. We have also addressed the admissibility of the proposed cost function and stability of the corresponding algorithm. The blind adaptation of the derived algorithm is shown to possess better convergence behavior compared to two existing algorithms. Finally, hints are provided to obtain blind equalization cost functions for square and cross quadrature amplitude modulation signals.

8. Exercises

1. Refer to Fig. 8 for geometrical details of square- and cross-QAM signals. Now following the ideas presented in Section 5, show that the blind equalization cost functions for square- and cross-QAM signals are respectively as follows:

$$\max_{w} \mathsf{E}\left[|y_n|^2\right], \text{ s.t. } \max\{|y_{R,n}|, |y_{I,n}|\} \leq R. \tag{89}$$

and

$$\max_{w} \mathsf{E}\left[|y_n|^2\right], \text{ s.t. } \max\{|\bullet\, y_{R,n}|, |y_{I,n}|\} + \max\{|y_{R,n}|, |\bullet\, y_{I,n}|\} - \max\{|y_{R,n}|, |y_{I,n}|\} \leq \bullet\, R. \tag{90}$$

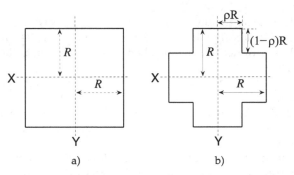

Fig. 8. Geometry of a) square- and b) cross-QAM $\left(\bullet = \frac{2}{3}\right)$.

2. By exploiting the independence between the in-phase and quadrature components of square QAM signal, show that the following blind equalization cost function may be obtained:

$$\max_{w} \mathsf{E}\left[|y_n|^2\right], \text{ s.t. } \max\{|y_{R,n}|\} = \max\{|y_{I,n}|\} \leq R. \tag{91}$$

The cost (91) originally appeared in Satorius & Mulligan (1993). Refer to Meng et al. (2009) and Abrar & Nandi (2010a), respectively, for its block-iterative and adaptive optimization.

9. Acknowledgment

The authors acknowledge the support of COMSATS Institute of Information Technology, Islamabad, Pakistan, King Fahd University of Petroleum and Minerals, Dhahran, Saudi Arabia, and the University of Liverpool, UK towards the accomplishment of this work.

10. References

Abrar, S. & Nandi, A.K. (2010a). Adaptive solution for blind equalization and carrier-phase recovery of square-QAM, *IEEE Sig. Processing Lett.* 17(9): 791–794.

Abrar, S. & Nandi, A.K. (2010b). Adaptive minimum entropy equalization algorithm, *IEEE Commun. Lett.* 14(10): 966–968.

Abrar, S. & Nandi, A.K. (2010c). An adaptive constant modulus blind equalization algorithm and its stochastic stability analysis, *IEEE Sig. Processing Lett.* 17(1): 55–58.

Abrar, S. & Nandi, A.K. (2010d). A blind equalization of square-QAM signals: a multimodulus approach, *IEEE Trans. Commun.* 58(6): 1674–1685.

Abrar, S. & Shah, S. (2006a). New multimodulus blind equalization algorithm with relaxation, *IEEE Sig. Processing Lett.* 13(7): 425–428.

Abrar, S. & Qureshi, I.M. (2006b). Blind equalization of cross-QAM signals, *IEEE Sig. Processing Lett.* 13(12): 745–748.

Abrar, S., Zerguine, A. & Deriche, M. (2005). Soft constraint satisfaction multimodulus blind equalization algorithms, *IEEE Sig. Processing Lett.* 12(9): 637–640.

Allen, J. & Mazo, J. (1973). Comparison of some cost functions for automatic equalization, *IEEE Trans. Commun.* 21(3): 233–237.

Allen, J. & Mazo, J. (1974). A decision-free equalization scheme for minimum-phase channels, *IEEE Trans. Commun.* 22(10): 1732–1733.

Baykal, B., Tanrikulu, O., Constantinides, A. & Chambers, J. (1999). A new family of blind adaptive equalization algorithms, *IEEE Sig. Processing Lett.* 3(4): 109–110.

Bellini, S. (1986). Bussgang techniques for blind equalization, *Proc. IEEE GLOBECOM'86* pp. 1634–1640.

Bellini, S. (1994). *Bussgang techniques for blind deconvolution and equalization*, in Blind deconvolution, S. Haykin (Ed.), Prentice Hall, pp. 8–59.

Benedetto, F., Giunta, G. & Vandendorpe, L. (2008). A blind equalization algorithm based on minimization of normalized variance for DS/CDMA communications, *IEEE Trans. Veh. Tech.* 57(6): 3453–3461.

Benveniste, A. & Goursat, M. (1984). Blind equalizers, *IEEE Trans. Commun.* 32(8): 871–883.

Benveniste, A., Goursat, M. & Ruget, G. (1980a). Analysis of stochastic approximation schemes with discontinous and dependent forcing terms with applications to data communications algorithms, *IEEE Trans. Automat. Contr.* 25(12): 1042–1058.

Benveniste, A., Goursat, M. & Ruget, G. (1980b). Robust identification of a nonminimum phase system: Blind adjustment of a linear equalizer in data communication, *IEEE Trans. Automat. Contr.* 25(3): 385–399.

Claerbout, J. (1977). Parsimonious deconvolution, *SEP-13* .

C.R. Johnson, Jr., Schniter, P., Endres, T., Behm, J., Brown, D. & Casas, R. (1998). Blind equalization using the constant modulus criterion: A review, *Proc. IEEE* 86(10): 1927–1950.

Ding, Z. & Li, Y. (2001). *Blind Equalization and Identification*, Marcel Dekker Inc., New York.

Donoho, D. (1980). On minimum entropy deconvolution, *Proc. 2nd Applied Time Series Symp.* pp. 565–608.

Farhang-Boroujeny, B. (1998). *Adaptive Filters*, John Wiley & Sons.

Fiori, S. (2001). A contribution to (neuromorphic) blind deconvolution by flexible approximated Bayesian estimation, *Signal Processing* 81: 2131–2153.

Garth, L., Yang, J. & Werner, J.-J. (1998). An introduction to blind equalization, *ETSI/STS Technical Committee TM6, Madrid, Spain* pp. TD7: 1–16.

Godard, D. (1980). Self-recovering equalization and carrier tracking in two-dimensional data communications systems, *IEEE Trans. Commun.* 28(11): 1867–1875.

Goupil, A. & Palicot, J. (2007). New algorithms for blind equalization: the constant norm algorithm family, *IEEE Trans. Sig. Processing* 55(4): 1436–1444.

Gray, W. (1978). Variable norm deconvolution, *SEP-14* .

Gray, W. (1979a). A theory for variable norm deconvolution, *SEP-15* .

Gray, W. (1979b). *Variable Norm Deconvolution*, PhD thesis, Stanford Univ.

Haykin, S. (1994). *Blind deconvolution*, Prentice Hall.

Haykin, S. (1996). *Adaptive Filtering Theory*, Prentice-Hall.

Im, G.-H., Park, C.-J. & Won, H.-C. (2001). A blind equalization with the sign algorithm for broadband access, *IEEE Commun. Lett.* 5(2): 70–72.

Kennedy, R. & Ding, Z. (1992). Blind adaptive equalizers for quadrature amplitude modulated communication systems based on convex cost functions, *Opt. Eng.* 31(6): 1189–1199.

Lucky, R. (1965). Automatic equalization for digital communication, *The Bell Systems Technical Journal* XLIV(4): 547–588.

Lucky, R. (1966). Techniques for adaptive equalization of digital communication systems, *The Bell Systems Technical Journal* pp. 255–286.

Meng, C., Tuqan, J. & Ding, Z. (2009). A quadratic programming approach to blind equalization and signal separation, *IEEE Trans. Sig. Processing* 57(6): 2232–2244.

Nikias, C. & Petropulu, A. (1993). *Higher-order spectra analysis a nonlinear signal processing framework*, Englewood Cliffs, NJ: Prentice-Hall.

Ooe, M. & Ulrych, T. (1979). Minimum entropy deconvolution with an exponential transformation, *Geophysical Prospecting* 27: 458–473.

Picchi, G. & Prati, G. (1987). Blind equalization and carrier recovery using a 'stop-and-go' decision-directed algorithm, *IEEE Trans. Commun.* 35(9): 877–887.

Pinchas, M. & Bobrovsky, B.Z. (2006). A maximum entropy approach for blind deconvolution, *Sig. Processing* 86(10): 2913–2931.

Pinchas, M. & Bobrovsky, B.Z. (2007). A novel HOS approach for blind channel equalization, *IEEE Trans. Wireless Commun.* 6(3): 875–886.

Romano, J., Attux, R., Cavalcante, C. & Suyama, R. (2011). *Unsupervised Signal Processing: Channel Equalization and Source Separation*, CRC Press Inc.

Rupp, M. & Sayed, A.H. (2000). On the convergence analysis of blind adaptive equalizers for constant modulus signals, *IEEE Trans. Commun.* 48(5): 795–803.

Sato, Y. (1975). A method of self-recovering equalization for multilevel amplitude modulation systems, *IEEE Trans. Commun.* 23(6): 679–682.

Satorius, E. & Mulligan, J. (1992). Minimum entropy deconvolution and blind equalisation, *IEE Electronics Lett.* 28(16): 1534–1535.

Satorius, E. & Mulligan, J. (1993). An alternative methodology for blind equalization, *Dig. Sig. Process.: A Review Jnl.* 3(3): 199–209.

Serra, J. & Esteves, N. (1984). A blind equalization algorithm without decision, *Proc. IEEE ICASSP* 9(1): 475–478.

Shalvi, O. & Weinstein, E. (1990). New criteria for blind equalization of non-minimum phase systems, *IEEE Trans. Inf. Theory* 36(2): 312–321.

Shtrom, V. & Fan, H. (1998). New class of zero-forcing cost functions in blind equalization, *IEEE Trans. Signal Processing* 46(10): 2674.

Sidak, Z., Sen, P. & Hajek, J. (1999). *Theory of Rank Tests*, Academic Press; 2/e.

Treichler, J. & Agee, B. (1983). A new approach to multipath correction of constant modulus signals, *IEEE Trans. Acoust. Speech Signal Processing* 31(2): 459–471.

Treichler, J. R. & Larimore, M. G. (1985). New processing techniques based on the constant modulus adaptive algorithm, *IEEE Trans. Acoust., Speech, Sig. Process.* ASSP-33(2): 420–431.

Tugnait, J.K., Shalvi, O. & Weinstein, E. (1992). Comments on 'New criteria for blind deconvolution of nonminimum phase systems (channels)' [and reply], *IEEE Trans. Inf. Theory* 38(1): 210–213.

Walden, A. (1985). Non-Gaussian reflectivity, entropy, and deconvolution, *Geophysics* 50(12): 2862–2888.

Walden, A. (1988). A comparison of stochastic gradient and minimum entropy deconvolution algorithms, *Signal Processing* 15: 203–211.

Wesolowski, K. (1987). Self-recovering adaptive equalization algorithms for digital radio and voiceband data modems, *Proc. European Conf. Circuit Theory and Design* pp. 19–24.

Widrow, B. & Hoff, M.E. (1960). Adaptive switching circuits, *Proc. IRE WESCON Conf. Rec.* pp. 96–104.

Widrow, B., McCool, J. & Ball, M. (1975). The complex LMS algorithm, *Proc. IEEE* 63(4): 719–720.

Wiggins, R. (1977). Minimum entropy deconvolution, *Proc. Int. Symp. Computer Aided Seismic Analysis and Discrimination* .

Wiggins, R. (1978). Minimum entropy deconvolution, *Geoexploration* 16: 21–35.

Yang, J., Werner, J.-J. & Dumont, G. (2002). The multimodulus blind equalization and its generalized algorithms, *IEEE Jr. Sel. Areas Commun.* 20(5): 997–1015.

Yuan, J.-T. & Lin, T.-C. (2010). Equalization and carrier phase recovery of CMA and MMA in blind adaptive receivers, *IEEE Trans. Sig. Processing* 58(6): 3206–3217.

Yuan, J.-T. & Tsai, K.-D. (2005). Analysis of the multimodulus blind equalization algorithm in QAM communication systems, *IEEE Trans. Commun.* 53(9): 1427–1431.

5

Adaptive Modulation for OFDM System Using Fuzzy Logic Interface

Seshadri K. Sastry
Dept. of Electronics and Communication Engineering
EIILM University, Sikkim
India

1. Introduction

Orthogonal Frequency Division Multiplexing (OFDM) is a multicarrier transmission technique, which divides the available spectrum into many carriers, each one being modulated by a low rate data stream. OFDM is a combination of modulation and multiplexing OFDM is a special case of Frequency Division Multiplexing (FDM), multiple user access is achieved by subdividing the available bandwidth into multiple channels that are then allocated to users.

However, OFDM uses the spectrum much more efficiently by spacing the channels much closer together. This is achieved by making all the carriers orthogonal to one another, preventing interference between the closely spaced carriers. Each carrier in an OFDM signal has a very narrow bandwidth (i.e. 1 kHz), thus the resulting symbol rate is low. This results in the signal having a high tolerance to multipath delay spread. One of the main reasons to use OFDM is to increase the robustness against frequency selective fading or narrowband interference. In a single carrier system, a single fade or interferer can cause the entire link to fail, but in a multicarrier system, only a small percentage of the subcarriers will be affected. Coded Orthogonal Frequency Division Multiplexing (COFDM) is the same as OFDM except that forward error correction is applied to the signal before transmission. This is to overcome errors in the transmission due to lost carriers from frequency selective fading, channel noise and other propagation effects.

OFDM overcomes most of the problems with both FDMA and TDMA. In FDMA many carriers are spaced apart in such a way that the signals can be received using conventional filters and demodulators. In such receivers, guard bands are introduced between carriers which results in lowering of spectrum efficiency. In OFDM sub carriers are mathematically orthogonal so that it is possible to receive signal without intercarrier interference. OFDM makes efficient use of spectrum by allowing overlap. OFDM is more resistant to frequency selective fading than single carrier systems due to dividing the channel into narrowband flat fading subchannels. OFDM eliminates ISI and IFI through use of a cyclic prefix. OFDM provides good protection against cochannel interference and impulsive parasitic noise. OFDM is less sensitive to sample timing offsets than single carrier systems (Seshadri Sastry et al., 2010 a) and (Seshadri Sastry et al., 2010 b) proposed a OFDM system with adaptive modulation using fuzzy logic interface to improve system capacity with maintaining good error performance. Adaptive modulation systems using ordinary hardware decision making circuits are inefficient to decide or change modulation scheme according to given conditions. Using fuzzy logic in decision making interface makes the system more efficient. The results of computer simulation show the improvement of system capacity in Rayleigh fading channel.

(Kwang et al. 2009) proposed a multi-user multiple-input multiple-output (MIMO) orthogonal frequency division multiplexing (OFDM) system with adaptive modulation and coding to improve system capacity with maintaining good error performance. The results of computer simulation show the improvement of system capacity in Rayleigh fading channel.

(Li Yanxin et al.2007) presented a novel method for demodulating the QAM signals basing on adaptive filtering. The commonly used least mean square (LMS) error adaptive filtering algorithm is employed for studying the demodulating procedure and the performance of the novel adaptive QAM demodulation. The novel adaptive QAM demodulation does not need the adaptive filter completing convergence. Therefore, the sampling rate and processing speed are decelerated. The performance of the method in theory is compared with computer simulating results. It shows that the error rates in simulation agree well with that in theory. Also, it is indicated that the demodulation method has many advantages over conventional ones, such as the powerful anti-noise ability, the small transfer delay, and the convenient implementation with DSP technology.

(Kiyoshi Hamaguchi et al.) proposed an adaptive modulation system for land mobile communications that can select one of quadrature amplitude modulation levels as a suitable modulation for propagation conditions is described. The main characteristics of the system are a mode in which information cannot be transmitted under adverse propagation conditions and a buffer memory for maintaining the data transmission rate. In the paper they confirmed that the basic performances of the adaptive modulation system using the equipment they developed and they found the measured performance was consistent with computer simulation results. Further in paper it was also confirmed that the adaptive modulation system provided a noticeable improvement in spectral efficiency and transmission quality.

Sorour Falahati, Arne Svensson, Torbjörn Ekman and Mikael Sternad proposed that when adaptive modulation is used to counter short – term fading in mobile radio channels, signaling delays create problems with outdated channel state information. The use of channel power prediction will improve the performance of the link adaptation. It is then of interest to take the quality of these predictions into account explicitly when designing an adaptive modulation scheme. They studied the optimum design of an adaptive modulation scheme based on uncoded M-QAM modulation assisted by channel prediction for the flat Rayleigh fading channel. The data rate, and in some variants the transmit power, are adapted to maximize spectral efficiency subject to average power and bit error rate constraints. The key issues studied here are how a known prediction error variance will affect the optimized transmission properties such as the SNR boundaries that determine when to apply different modulation rates, and to what extent it affects the spectral efficiency. The investigation is performed by analytical optimization of the link adaptation, using the statistical properties of a particular but efficient channel power predictor.

Optimum solutions for the rate and transmit power are derived based on the predicted SNR and the prediction error variance.

M.K.Wasantha and W.A.C.Fernando discussed an OFDM-CDMA system with adaptive modulation schemes for future generation wireless networks. Results presented in this paper show that adaptive systems can perform better than fixed modulation based systems both in terms of BER and spectral efficiency.

2. OFDM generation

In OFDM we have N subcarriers, N can be anywhere from 16 to 1024 in present technology and depends on environment it is used. Block diagram of OFDM system is shown in Fig 1 below. OFDM transmitter consists of Serial to parallel converter, modulator, IFFT block, parallel to serial converter and block to add cyclic prefix. OFDM receiver consists of block to remove cyclic prefix, Serial to parallel converter, IFFT block, de modulator and parallel to serial converter

2.1 OFDM transmitter

Figure 1 shows the setup for a basic OFDM transmitter and receiver. An OFDM transmitter converts serial data to parallel, modulates it, converts it to time domain and transmits serial data.

2.1.1 Serial to parallel conversion

The input serial data stream is formatted into the word size required for transmission, e.g. 2bit/word for QPSK, and shifted into a parallel format. The data is then transmitted in parallel by assigning each data word to one carrier in the transmission.

2.1.2 Modulator

Modulation is a process of facilitating the transfer of information over a medium. Modulation is the process of mapping of the information on changes in carrier phase, frequency or amplitude or combination. Modulation schemes such as PSK (BPSK, QPSK, 8-PSK, 16-PSK, 32-PSK) or QAM (8-QAM, 16-QAM, 32-QAM, 64-QAM) are used.

2.1.3 Inverse fast fourier transform (IFFT)

IFFT block is used to change domain of the signal from frequency to time. IFFT is a mathematical concept which accepts amplitudes of some sinusoids; crunch these numbers to produce time domain result. Both IFFT and FFT will produce identical result on same input

2.1.4 Adding cyclic prefix

Adding cyclic prefix is a process in which we extend the symbol such each symbol is more than one cycle, which allows the symbol to be out of delay spread zone and it is not corrupted. Cycli prefix will be around 10% - 25% of the symbol time. Addition of cyclic prefix mitigates the effects of fading, intersymbol interference and increases bandwidth..

2.2 Channel

A channel model is then applied to the transmitted signal. The model allows for the signal to noise ratio, multipath, and peak power clipping to be controlled. The signal to noise ratio is set by adding a known amount of white noise to the transmitted signal. Multipath delay spread then added by simulating the delay spread using an FIR filter. The length of the FIR filter represents the maximum delay spread, while the coefficient amplitude represents the reflected signal magnitude.

2.3 Receiver

The receiver basically does the reverse operation to the transmitter. The guard period is removed. The FFT of each symbol is then taken to find the original transmitted spectrum. The phase angle of each transmission carrier is then evaluated and converted back to the data word by demodulating the received phase. The data words are then combined back to the same word size as the original data.

2.3.1 Removing cyclic prefix

Cyclic prefix added in the transmitter is removed to get perfect periodic signal

2.3.2 FFT

FFT block performs reverse operation to IFFT block. It is used to change domain of the signal from time to frequency

2.3.3 De Modulator

A demodulator performs reverse process of modulator which returns information

2.3.4 Parallel to serial converter

Using parallel to serial converter

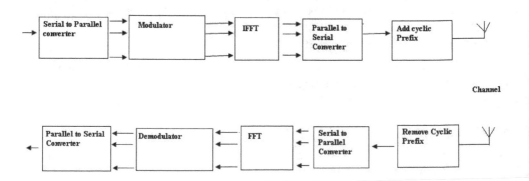

Fig. 1. OFDM system

3. Implementing OFDM system

An OFDM system was modeled using Matlab to allow various parameters of the system to be varied and tested. The aim of doing the simulations was to measure the performance of OFDM under different channel conditions, and to allow for different OFDM configurations to be tested. The main criterion of this chapter is to compare performance of Fuzzy logic based adaptive modulated OFDM system with adaptive modulated OFDM system. Matlab program to implement OFDM system and simulation results are given below

3.1 Matlab program to implement OFDM system

3.1.1 Transmitter design

3.1.1.1 Defining parameters

M = 16; % Size of signal constellation
k = log2(M); % Number of bits per symbol
n = 3e4; % Number of bits to process
nsamp = 1; % Oversampling rate

3.1.1.2 Signal source

t_data = randint(9600,1); % Random binary data stream
% Plot first 40 bits in a stem plot.
stem(t_data(1:40),'filled');
title('Random Bits');
xlabel('Bit Index'); ylabel('Binary Value');

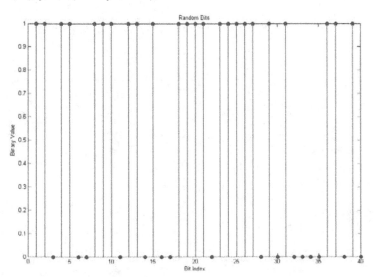

Fig. 2. Random Bits stream (Used by signal source)

xsym = bi2de(reshape(t_data,k,length(t_data)/k).','left-msb');.

```
figure;.
stem(xsym(1:10));
title('Random Symbols');
xlabel('Symbol Index'); ylabel('Integer Value');
close all
clear all
clc
M = 16; % Size of signal constellation
k = log2(M); % Number of bits per symbol
t_data=randint(9600,1)';
stem(t_data(1:40),'filled');
title('Random Bits');
xlabel('Bit Index'); ylabel('Binary Value');
ti=(0:1:49);
```

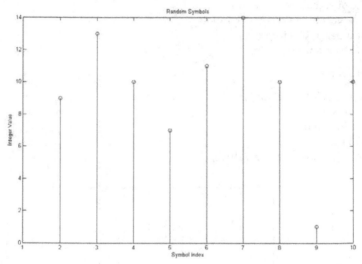

Fig. 3. Bits converted to symbols

```
xsym = bi2de(reshape(t_data,k,length(t_data)/k).','left-msb');
%% Stem Plot of Symbols
% Plot first 10 symbols in a stem plot.
figure; % Create new figure window.
stem(xsym(1:10));
title('Random Symbols');
xlabel('Symbol Index'); ylabel('Integer Value');
x=1;
si=1;
for d=1:100;
data=t_data(x:x+95);
x=x+96;
k=3;
```

```
n=6;
s1=size(data,2); % Size of input matrix
j=s1/k;
```

3.1.1.3 Convolutional encoding

```
constlen=7;
codegen = [171 133]; % Polynomial
trellis = poly2trellis(constlen, codegen);
codedata = convenc(data, trellis);
```

3.1.1.4 Interleaving data

```
s2=size(codedata,2);
j=s2/4;
matrix=reshape(codedata,j,4);
intlvddata = matintrlv(matrix',2,2)'; % Interleave.
intlvddata=intlvddata';
 dec=bi2de(intlvddata','left-msb');
```

3.1.1.5 16-QAM modulation

```
M=16;
y = qammod(dec,M);
```

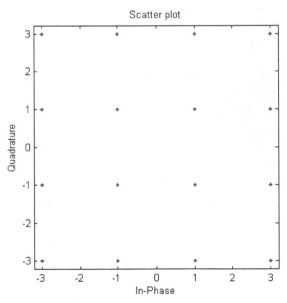

Fig. 4. Constellation ordering of 16-QAM modulator

3.1.1.6 Pilot insertion

```
lendata=length(y);
pilt=3+3j;
```

```
nofpits=4;
k=1;
for i=(1:13:52)
 pilt_data1(i)=pilt;
  for j=(i+1:i+12);
 pilt_data1(j)=y(k);
 k=k+1;
  end
end
pilt_data1=pilt_data1'; % size of pilt_data =52
pilt_data(1:52)=pilt_data1(1:52); % upsizing to 64
pilt_data(13:64)=pilt_data1(1:52); % upsizing to 64
for i=1:52
 pilt_data(i+6)=pilt_data1(i);
end
3.8 IFFT
ifft_sig=ifft(pilt_data',64);
```

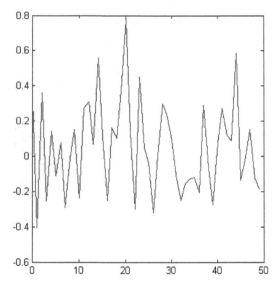

Fig. 5. OFDM Signal

3.1.1.7 Adding cyclic extension

```
cext_data=zeros(80,1);
cext_data(1:16)=ifft_sig(49:64);
for i=1:64

 cext_data(i+16)=ifft_sig(i);

end
```

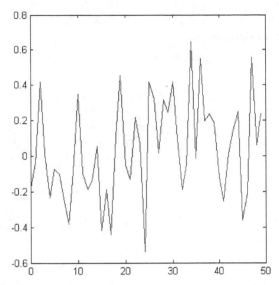

Fig. 6. OFDM Signal after Cyclic prefix

3.1.1.8 Channel

o=1;
for snr=0:2:50
ofdm_sig=awgn(cext_data,snr,'measured'); % Adding white Gaussian Noise

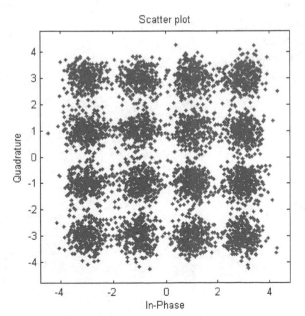

Fig. 7. Constellation ordering at receiver

3.1.2 Receiver design

3.1.2.1 Removing cyclic extension

```
for i=1:64
 rxed_sig(i)=ofdm_sig(i+16);
end
% FFT
ff_sig=fft(rxed_sig,64);
% Pilot Synch
for i=1:52
 synched_sig1(i)=ff_sig(i+6);
end
k=1;
for i=(1:13:52)
 for j=(i+1:i+12);
 synched_sig(k)=synched_sig1(j);
 k=k+1;
 end
end
```

3.1.2.2 Demodulation

```
dem_data= qamdemod(synched_sig,16);
bin=de2bi(dem_data','left-msb');
bin=bin';
```

3.1.2.3 De-interleaving

```
deintlvddata = matdeintrlv(bin,2,2); % De-Interleave
deintlvddata=deintlvddata';
deintlvddata=deintlvddata(:)';
```

3.1.2.4 Decoding data

```
n=6;
k=3;
decodedata  =vitdec(deintlvddata,trellis,5,'trunc','hard'); % decoding datausing veterbi
decoder
rxed_data=decodedata;
```

3.1.2.5 Calculating BER

```
rxed_data=rxed_data(:)';
errors=0;
c=xor(data,rxed_data);
errors=nnz(c);
BER(si,o)=errors/length(data);
o=o+1;
 end si=si+1;
end for col=1:25;
 ber(1,col)=0;
```

```
for row=1:100;
 ber(1,col)=ber(1,col)+BER(row,col);
 end
end
ber=ber./100;
%%
figure
i=0:1:49;
semilogy(i,ber);
title('BER vs SNR');
ylabel('BER');
xlabel('SNR (dB)');
grid on
```

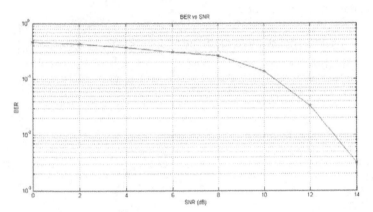

Fig. 8. Bit error rate of OFDM system

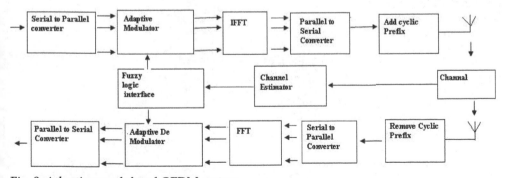

Fig. 9. Adaptive modulated OFDM system

4. Adaptive modulation for OFDM system using fuzzy logic interface

In a OFDM system using lower order modulators such asBPSK, 4 QAM and 8 QAM will improve Bit Error Rate (BER) but decreases spectral efficiency and speed, on the other hand employing higher order modulators such as 64 QAM, 128 QAM, 256QAM and 512 QAM will increase spectral efficieny and speed but result in poor BER. So to achieve good trade-off between spectral efficiency and overall BER (Bit Error Rate). Adaptive modulation is used.

4.1 Adaptive modulator and demodulator

At the transmitter the adaptive modulator block consists of different modulators which are used to provide different modulation orders. The switching between these modulators will depend on the instantaneous SNR. The goal of adaptive modulation is to choose the appropriate modulation mode for transmission depending on instantaneous SNR, in order to achieve good trade-off between spectral efficiency and overall BER. Adaptive modulation is a powerful technique for maximizing the data throughput of subcarriers allocated to a user. Adaptive modulation involves measuring the SNR of each subcarrier in the transmission, then selecting a modulation scheme that will maximize the spectral efficiency, while maintaining an acceptable BER.

4.2 Program to implement adaptive modulation for OFDM

4.2.1 Defining parameters

```
clear
 N = 256;
 P = 256/8;
 S = N-P;
 GI = N/4;
 M = 2;
 pilotInterval = 8;
 L = 16;
 nIteration = 500;
 SNR_V = [0:1:34];
 ber = zeros(1,length(SNR_V));
 Ip = [1:pilotInterval:N];
 Is = setxor(1:N,Ip);
 Ep = 2;
```

4.2.2 fft

```
F = exp(2*pi*sqrt(-1)/N .* meshgrid([0:N-1],[0:N-1])...
.* repmat([0:N-1]',[1,N]));
for( i = 1 : length(SNR_V))
SNR = SNR_V(i)
if (SNR>=0 && SNR<=2)
M=2;
elseif (SNR>2 && SNR<=4)
M=4;
elseif (SNR>4 && SNR<=8)
M=8;
elseif (SNR>8 && SNR<=14)
M=16;
```

```
elseif (SNR>14 && SNR<=20)
M=32;
elseif (SNR>20 && SNR<=27)
M=64;
elseif (SNR>27 && SNR<=34)
M=128;
end
for(k = 1 : nIteration)
h(1:L,1) = random('Normal',0,1,L,1) + ...
j * random('Normal',0,1,L,1);
h = h./sum(abs(h));
```

4.2.3 Transmission of data

```
TrDataBit = randint(N,1,M);
TrDataMod = qammod(TrDataBit,M);
TrDataMod(Ip) = Ep * TrDataMod(Ip);
TrDataIfft = ifft(TrDataMod,N);
TrDataIfftGi = [TrDataIfft(N- GI + 1 : N);TrDataIfft];
TxDataIfftGi = filter(h,1,TrDataIfftGi);
TxDataIfftGiNoise = awgn(TxDataIfftGi ...
, SNR - db(std(TxDataIfftGi)));
TxDataIfft = TxDataIfftGiNoise(GI+1:N+GI);
TxDataMod = fft(TxDataIfft,N);
```

4.2.4 Channel estimation

```
Spilot = TrDataMod(Ip);
Ypilot = TxDataMod(Ip);
G = (Ep * length(Ip))^-1 ...
* ctranspose(sqrt(Ep)*diag(Spilot)*ctranspose(F(1:L,Ip)));
hHat = G*Ypilot;
TxDataBit = qamdemod(TxDataMod./(fft(hHat,N)),M);
```

4.2.5 Bit error rate computation

```
[nErr bErr(i,k)] = symerr(TxDataBit(Is),TrDataBit(Is));
end
end
f1 = figure(1);
set(f1,'color',[1 1 1]);
semilogy(SNR_V,mean(bErr'),'r-d')
xlabel('SNR ');
ylabel('BER')
grid on;
hold on;
```

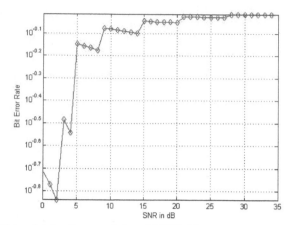

Fig. 10. Bit Error Rate of Adaptive modulated OFDM System

5. FIS (Fuzzy Interface system)

FIS (Fuzzy Interface system) is the decision making system in Channel Estimator (SNR estimator) used in adaptive modulation. It is modeled in Matlab 7.4 Fuzzy Interface editor. It takes instantaneous SNR (Signal to Noise Ratio) and Present modulation order as inputs and controls the modulation order of modulator and demodulator blocks.

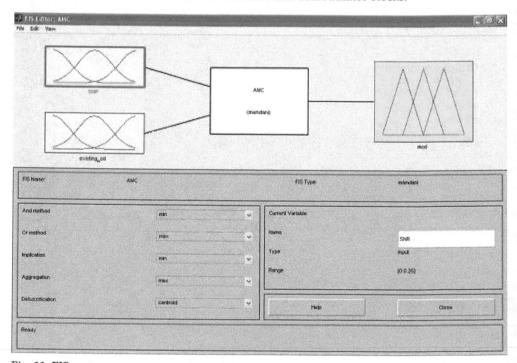

Fig. 11. FIS system

FIS (Fuzzy Interface System) consists of two inputs and one output as shown in figure above Input one SNR and input two present mod. The membership function of SNR and present mod is shown in figures 12 and 13. For SNR, three membership functions are taken namely low_SNR, medium_SNR and high_SNR. For present modulation six membership functions are taken namely QPSK, 8QAM, 16QAM, 32QAM, and 64QAM and128QAM.

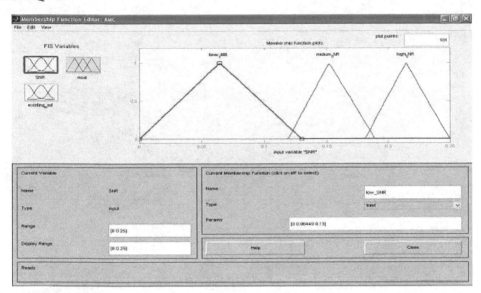

Fig. 12. SNR membership functions

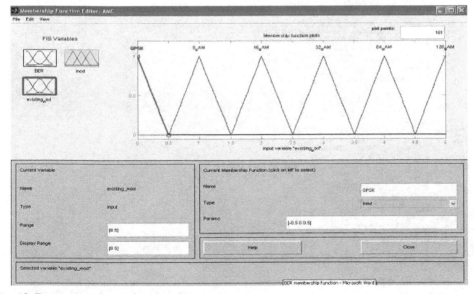

Fig. 13. Present mod membership functions

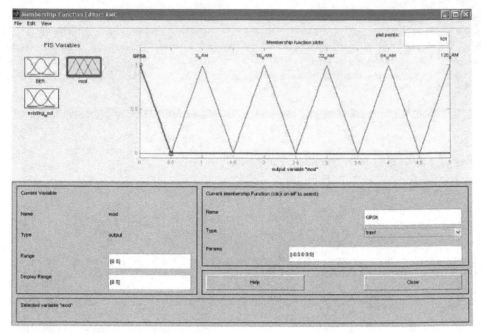

Fig. 14. Output membership function

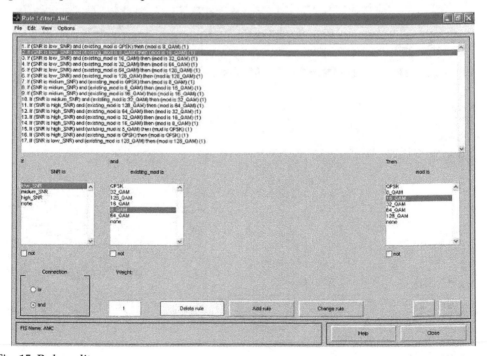

Fig. 15. Rules editor

Rules are edited in rules editor which gives conditions to select modulation order depending on channel estimation (SNR). Example for rules are as follows , if BER is low increase the modulation order by one level compared to present modulation order , if BER is high then decrease the modulation order by one compared to present modulation order. If BER is average then increase the modulation order by one compared to present modulation order.

The above proposed system was simulated in Matlab7.4., Using fuzzy logic in decision making is a good choice because ordinary (non fuzzy) system is controlled by plain if and else statements, for example, if for poor SNR (Signal To Noise Ratio) range is declared as 0 to 4 , if input is 4.1 then the input is not considered as poor SNR (But it is poor). If we use fuzzy logic in above case 4.1 is also considered as poor SNR. So using FIS (Fuzzy interface system) increases the performance adaptive modulation system.

Fig 16 shows the output of FIS (fuzzy interface system) for given set of inputs, output is selected based on given rules. Bit Error Rate performance of the simulated system is shown in Fig 17.

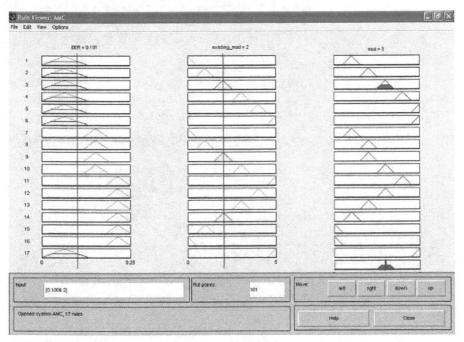

Fig. 16. Simulation result

Fig 17 shows comparison bit error rates of adaptive modulated OFDM system and fixed modulated OFDM system. Fig 18 gives plot showing the comparison of performances of adaptive modulated OFDM system using fuzzy logic and adaptive modulated OFDM system using ordinary control logic. It was shown that OFDM system using Fuzzy logic performs better than OFDM system using ordinary control logic. Using FIS (Fuzzy interface system) in implementing adaptive modulation for OFDM system increases performance of system since it responds to channel condition and maintains good performance (Bit Error Rate) and capacity (spectral efficiency) efficiently than system using ordinary control logic.

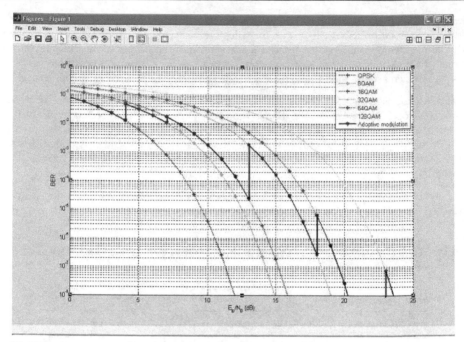

Fig. 17. BER comparison of proposed scheme and fixed modulation schemes.

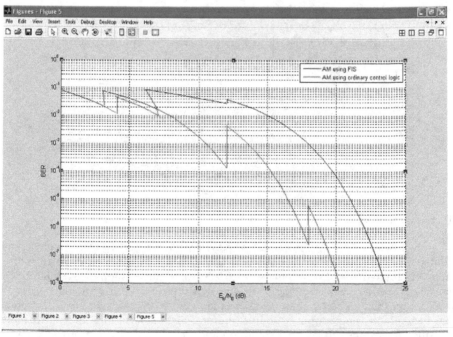

Fig. 18. Comparing Adaptive modulation schemes using FIS and using ordinary control logic

6. Conclusion

Adaptive modulation is a powerful technique for maximizing the data throughput of subcarriers allocated to a user. Adaptive modulation involves measuring the SNR of each subcarrier in the transmission, then selecting a modulation scheme that will maximize the spectral efficiency, while maintaining an acceptable BER. Fuzzy logic based adaptive modulation performs better ordinary logic based adaptive modulation for OFDM system.

7. References

K. Seshadri Sastry and Dr.M.S.Prasad Babu, "Adaptive Modulation for OFDM system using Fuzzy logic interface", july 15-18, 2010, IEEE ICSESS, PP 368-371.

K. Seshadri Sastry and Dr.M.S.Prasad Babu, "Fuzzy logic based Adaptive Modulation Using Non Data Aided SNR Estimation for OFDM system", International Journal of Engineering Science and Technology, Vol. 2(6), 2010, pp 2384-2392

Kwang Yoon Kim, Yoon Hyun Kim and Jin Young Kim "Performance of Multi-user MIMO OFDM System with Adaptive Modulation and coding for wireless communications" Feb 15-18, ICACT 2009 Modulation and Coding for Wireless Communications", feb 15-19, 2009, IEEE ICACT 2009

Li Yanxin, Hu Aiqun, "An Adaptive Demodulation Method for QAM Signals" IEEE 2007 International Symposium on Microwave, Antenna, Propagation, and EMC Technologies For Wireless Communications

K.Seshadri Sastry and Dr.M.S.Prasad Babu ", SNR Estimation for QAM signals Using Fuzzy Logic Interface", IEEE ICCSIT 2010, July 2010, pp 413- 416

Jind-Yeh Lee, Huan Chang and Henry Samueli, "A Digital Adaptive Beamforming QAM Demodulator IC for High Bit-Rate Wireless Communications", IEEE Journal of solid - state circuits, vol. 33, NO. 3, Mar 1998, pp 367-377.

Jouko Vankka, Marko Kosunen, Ignacio Sanchis, and Kari A. I. Halonen, "A Multicarrier QAM Modulator", IEEE Transactions on circuits and systems —II:Analog and digital signal processing, vol. 47, No. 1, jan 2000,pp 1-10.

Franc.ois-Xavier Socheleau, Abdeldjalil A˙issa-El-Bey, and S´ebastien Houcke, " Non Data-Aided SNR Estimation of OFDM Signals", IEEE Communication Letters, VOL. 12, NO. 11, NOV 2008, pp 813-815

K. Seshadri Sastry and Dr.M.S.Prasad Babu," AI based Digital Companding Scheme for OFDM system using custom constellation Mapping and selection", International Journal on Computer Science and Engineering , Vol. 02, No. 04, 2010, pp 1381-1386

Li Yanxin, Hu Aiqun, "An Adaptive Demodulation Method for QAM Signals" IEEE 2007 International Symposium on Microwave, Antenna, Propagation, and EMC Technologies For Wireless Communications

K. Seshadri Sastry and Dr.M.S.Prasad Babu," AI Based Digital Companding scheme for Software Defined Radio" IEEE ICSESS 2010, pp 417- 419

Kiyoshi Hamaguchi, Yukiyoshi Kamio and Eimatsu Moriyama, "Implementation and Performance of an Adaptive QAM Modulation level-controlled System for Land Mobile Communications"

Falahati, Arne Svensson, Torbjörn Ekman and Mikael Sternad, "Adaptive Modulation Systems for Predictive Wireless Channels"

K. Seshadri Sastry and Dr.M.S.Prasad Babu," Digital Companding Scheme using A I based custom constellation Mapping and selection

K. Pahlavan and J. L. Holsinger, "A Model for the Effects of PCM Compandors on the Performance of High Speed Modems," GLOBECOM '85, New Orleans, December 1985, pp. 24.8.1-24.8.5

M. K. Wasantha and W. A. C. Fernando, "QAM based Adaptive Modulation for OFDM-CDMA Wireless Networks",

Li Yanxin, Hu Aiqun, "An Adaptive Demodulation Method for QAM Signals" IEEE 2007 International Symposium on Microwave, Antenna, Propagation, and EMC Technologies For Wireless Communications",

S. Sasaki, J. Zhu and G. Marubayashi, "Performance of parallel combinatory spread spectrum multiple access communication systems," Proceedings of 1991 IEEE International Symposium on Personal, Indoor and Mobile Radio Communications (PIMRC), pp.204-208.

6

Security Limitations of Spectral Amplitude Coding Based on Modified Quadratic Congruence Code Systems

Hesham Abdullah Bakarman, Shabudin Shaari and P. Susthitha Menon
University Kebangsaan Malaysia
Malaysia

1. Introduction

Generally, communication network systems provide data transfer services for customers. Further requirements such as performance, security, and reliability characterize the quality of the transfer service. Network and information security refer to confidence that information and services existing on a network cannot be accessed by unauthorized users (eavesdropper). However, these service requirements affect each other such that a decision has to be made for cases in which all or some of these requirements are desired but cannot be fulfilled (Zorkadis 1994).

In secure communication networks, tradeoff considerations between system performance and security necessities have not been mentioned widely in many researches. Actually, it has been known that security is of main concern in both wireless and optical communications networks, security mechanisms employed often have implication on the performance of the system (Imai et al. 2005). For some application environments, such as military or enterprise networks, security and system capacity in communications transmission media could become a critical issue. Optical code-division multiple-access (optical CDMA) technology, a multiplexing technique adapted from the successful implementation in wireless networks, is an attractive solution for these applications because it presents security in the physical layer while providing significantly wide bandwidth (Chung et al. 2008).

Optical CDMA systems are getting more and more attractive in the field of all optical communications as multiple users can access the network asynchronously and simultaneously with high level of security (Salehi 1989, Salehi & Brackett 1989) compared to other multiplexing techniques such as Wavelength Division Multiplexing WDM and Time Division Multiplexing TDM.

The potential provided by optical CDMA for enhanced security is frequently mentioned in several studies using different techniques and approaches such as quantum cryptography and chaotic encryption systems (Castro et al. 2006). Other approaches to enhance security have been proposed using optical encoding techniques such as fiber bragg gratings (FBG) to implement optical CDMA systems (Shake 2005a,2005b). Their degree of security depends on code dimensions being used.

In this chapter, security limitations of spectral amplitude coding Optical CDMA are presented and investigated. The tradeoffs between security and system performance have been investigated for a specific eavesdropper interception situation. Section II briefly presents some network security services and assumptions required for optical CDMA confidentiality analysis in the physical layer. Security and performance tradeoffs, based on MQC code system, are presented in section III. Performance analysis is given in section IV. Finally, a conclusion is given in section V.

2. Optical CDMA physical layer networks

Due to the transparency increment in optical communications network components and systems, network management and maintenance have been faced additional security challenges. An evaluation on several existing physical security violates on optical communications network is presented in (Teixeira et al. 2008).There are four main threats that can be described in terms of how they affect the normal flow of information in the network, as shown in figure (1),they are: denial of service, interception, modification and creation. Table 1 summarized some of these attacks.

Attack method	Realizes	Means
In-Band Jamming	Service Disruption	An attacker injects a signal designed to reduce the ability of the receiver to interpret correctly the transmitted data
Out-of-Band Jamming	Service Disruption	An attacker reduces communication signal component by exploiting leaky components or cross-modulation effects
Unauthorized Observation	Eavesdropping	An attacker listens to the crosstalk leaking from an adjacent signal through a shared resource in order to gain information from the adjacent signal, the collection of signals by an attacker for whom they were not intended).

Table 1. Optical networks attack methods

The security services of a network have four fundamental objectives designed to protect the data and the network's resources (Fisch & White 2000). These objectives are:

- Confidentiality: ensuring that an unauthorized individual does not gain access to data contained on a resource of the network.
- Availability: ensuring that authorized users are not unduly denied access or use of any network access for which they are normally allowed.
- Integrity: ensuring that data is not altered by unauthorized individuals. Related to this is authenticity which is concerned with the unauthorized creation of data.
- Usage: ensuring that the resources of the network are reserved for use only by authorized users in appropriate manner.

In this chapter, ensuring confidentiality against eavesdropper interception strategies for optical CDMA aims to investigate the limitations and tradeoffs between security and performance.

There are various fiber optic tapping methods, of which fall into the following main categories (Oyster Optics 2008): splice (involves literally breaking the cable at some point and adding a splitter), splitter or coupler (involves bending the cable to a certain radius, which allows a small amount of the transmitted light to escape) and non-touching methods (passive and active), involve highly sensitive photo-detectors that capture the tiny amounts of light that emerge laterally from the glass fiber owing to a phenomenon known as Rayleigh scattering.

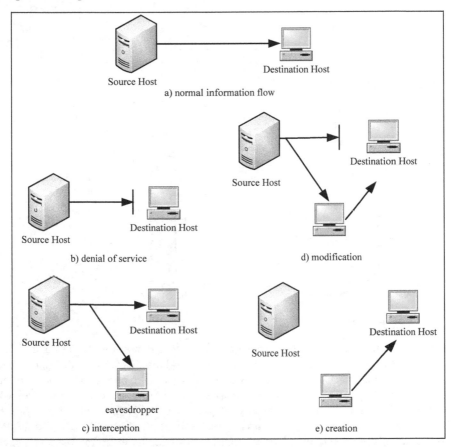

Fig. 1. Pattern of network attacks

Communication between authorized users in a network can be implemented by two approaches; point-to-point and broadcast. In the point-to-point, approach each user transmits to another specific one whereas in a broadcast approach users transmit in common to the medium accessible to all other users. Figure (2) shows a common topology found in point-to-point networks. Figure (3) shows two topologies established in broadcast networks.

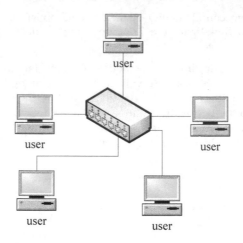

Fig. 2. Point to point star topology

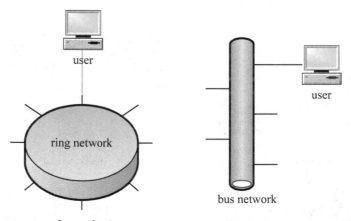

Fig. 3. Broadcast network topologies

Topology is an important architectural consideration, and different topologies have different security properties. Ring topologies allow attacks to be relatively easily localized, because of the structured interconnectivity of nodes. Star topologies make attack detection nominally easier than other topologies, because any propagating attacks are commonly received at many stations. Optical CDMA has many advantages such as sharing bandwidth, controlling and high security compared to other access technologies such as TDMA and WDMA. Recently, studies discovered that Optical CDMA systems suffer from weakness against eavesdropping and jamming attacks.

Figure (4) shows the possible positions, within the network, to tap a signal from the user. Therefore, when just a single user is active, optical CDMA system cannot guarantee physical layer security any more. In certain time, this situation can be existed even in a multiuser active optical CDMA network as reported in current theoretical analyses (Shake 2005a,2005b).

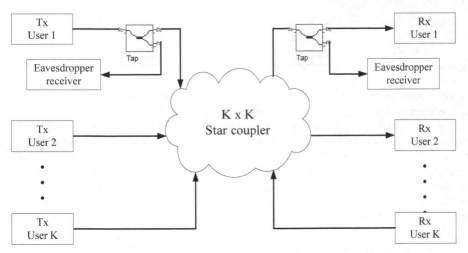

Fig. 4. Places for an eavesdropper to attack and tap optical CDMA encoded pulses

3. Security and performance tradeoffs

In security environments, it is believed that an inherent tradeoffs between networks performance and security are existed which lead many network designers to seek a balance between both of them. Depending on the confidentiality measurement required between communicating networks, different sets of optimizations can be considered (Jin-Hee & Ing-Ray 2005). In (Wolter & Reinecke 2010), the relationship of performance and security has been investigated in model-based evaluation. Their approach is illustrated based on the premise that there are significant similarities between security and reliability.

The combination of security and performance poses interesting tradeoffs that have high relevance especially in modern systems that are subject to requirements in areas, performance and security. In this chapter, ensuring confidentiality against eavesdropper interception strategies for optical CDMA is conducted to investigate limitations and tradeoffs between security and performance.

Using the modeling approximations of (Shake 2005b), per signature chip SNR of the eavesdropper is related to the per data bit signal-to-noise ratio (SNR) of the user by the following relationship:

$$\frac{E_{ed}}{N_{0ed}} = \sigma\left(\frac{1}{W}\right)\left(\frac{1}{1 - \dfrac{M_A}{M_T}}\right)\left(\frac{E_u}{N_{0u}}\right)_{spec} \tag{1}$$

W is the code weight of the code being used, M_T is the maximum theoretical number of simultaneous users at a specified maximum BER, E_u / N_{0u} is the required user SNR (per data bit) to maintain the specified BER, M_T is the actual number of simultaneous users supported, and E_{ed} / N_{0ed} is the eavesdropper's effective SNR per code chip. Where σ represents several system design parameters as following:

$$\sigma = \left(\frac{e_t n_u}{\alpha_{ed} e_u} \right) \qquad (2)$$

In this equation, e_t is the eavesdropper's fiber tapping efficiency, n_u is the number of taps in the broadcast star coupler that distributes user signals, α_{ed} is the ratio of the eavesdropper's receiver noise density to the authorized user's receiver noise density, e_u is the authorized user receiver's multichip energy combining efficiency. Figure (5) shows the effect of combining multiple code pulses for both coherent and incoherent detection schemes. The eavesdropper is assumed to use a receiver that is equal in sensitivity to the authorized user's receiver ($\alpha_{ed} = 1$). It is assumed that the total number of taps in the star coupler, shown in figure (4), is $n_u = 100$ with a tapping efficiency of $e_t = 0.01$. Since, e_u is equal to one and between zero and one for coherent and incoherent detection respectively (Mahafza & Elsherbeni 2003), coherent detection with combining signals shows better confidentiality than the incoherent one.

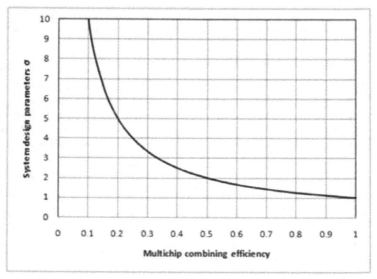

Fig. 5. Effect of combining multiple code pulses for both coherent and incoherent detection schemes

4. Performance analysis

Spectral amplitude coding optical CDMA systems using codes, which have code properties with low in-phase cross correlation, can eliminate the interference signals such as M-sequence (Peterson et al. 1995), Hadamard (Zou,Ghafouri-Shiraz et al. 2001), modified double weight (MDW) (Aljunid et al. 2004), and modified quadratic congruence (MQC) (Zou,Shalaby et al. 2001) codes. However, as broad-band thermal sources are used in such system, the phase-induced intensity noise (PIIN) that is due to the intensity fluctuation of thermal source severely affects the system performance (Smith et al. 1998). Commonly, these codes are represented by (N, w, λ) notation where N, w, and λ are code length, code weight, and in-phase cross correlation, respectively.

The establishment of MQC codes was proposed in (Zou,Shalaby et al. 2001). The proposed code families with the odd prime number p > 1 and represented by (p2+p, p+1, 1), have the following properties:

i.   there are p2 sequences.
ii.   each code sequence has N = (p2+p) chip component that can be splitted into w = (p+1) sets, and each set consists of one "1" and (p-1) "0 s".
iii. Between any two sequences cross correlation λ is exactly equal to 1.

According to (Zou,Shalaby et al. 2001), MQC code families can be constructed in two steps as following:

Step 1: Let GF (p) represents a finite field of p elements. A number sequence $y_{\alpha,\beta}(k)$ is assembled with elements of GF (p) over an odd prime by using the following expression:

$$y_{\alpha,\beta}(k) = \begin{cases} d[(k+\alpha)^2+\beta](\bmod p), k=0,1,...,p-1 \\ [\alpha+b](\bmod p), k=p \end{cases} \tag{3}$$

where d ∈ {0, 1, 2, ..., p-1} and b, α, β ∈ {0, 1, 2, ..., p-1}.

Step 2: a sequence of binary numbers $s_{\alpha,\beta}(i)$ is constructed based on each generated number sequence $y_{\alpha,\beta}(k)$ by using the following mapping method:

$$s_{\alpha,\beta}(i) = \begin{cases} 1, if i=kp+y_{\alpha,\beta}(k) \\ 0, otherwise \end{cases} \tag{4}$$

where i = 0, 1, 2, ..., p2+p-1, k = $\lfloor i/p \rfloor$. Here, $\lfloor x \rfloor$ defines the floor function of x.

Table 2 shows MQC basic code matrix for p = 3. Thus, the code length N = 12, code weight w = 4, and in-phase cross correlation is 1. The upper bound of the number of codes that can be produced is p2 = 9 code sequences.

In the analysis of spectral-amplitude coding system, PIIN, shot noise and thermal noise are three main noises that should be taken into consideration. To simplify the analysis, the distribution of intensity noise and shot noise are approximated as Gaussian for calculating the bit-error-rate (BER). The analysis performance of optical CDMA system based on MQC codes in the existence of PIIN, the photodiode shot noise and the thermal noise are presented in (Zou,Shalaby et al. 2001). Based on the complementary detection scheme the average signal to noise ratio has been expressed as:

$$SNR = \frac{I^2_{Data}}{\langle I^2_{Total\,noise} \rangle} \tag{5}$$

$$I^2_{Data} = \frac{\mathfrak{R}^2 P_{sr}^2}{p^2} \tag{6}$$

P_{sr} is the effective power of a broadband source at the receiver and $\mathfrak{R}$ is the photodiode responsivity.

And

$$\langle I^2_{Total\,noise} \rangle = \langle I^2_{shot} \rangle + \langle I^2_{PIIN} \rangle + \langle I^2_{thermal} \rangle \tag{7}$$

		C1	C2	C3	C4	C5	C6	C7	C8	C9	C10	C11	C12
						Code spectral chips							
Code sequences (users)	1	1	0	0	0	1	0	0	1	0	0	0	1
	2	0	1	0	0	0	1	0	0	1	0	0	1
	3	0	0	1	1	0	0	1	0	0	0	0	1
	4	0	1	0	0	1	0	1	0	0	1	0	0
	5	0	0	1	0	0	1	0	1	0	1	0	0
	6	1	0	0	1	0	0	0	0	1	1	0	0
	7	0	1	0	1	0	0	0	1	0	0	1	0
	8	0	0	1	0	1	0	0	0	1	0	1	0
	9	1	0	0	0	0	1	1	0	0	0	1	0

Table 2. MQC basic code matrix for p = 3

Then,

$$SNR = \frac{\frac{\Re^2 P_{sr}^2}{p^2}}{\frac{P_{sr}\,e\,B\,\Re}{N}[p-1+2K] + \frac{P_{sr}^2\,B\Re^2 K}{2\,\Delta\upsilon\,wp^2}\left[\frac{(K-1)}{p}+p+K\right] + \frac{4K_b T_n B}{R_L}} \tag{8}$$

Where e is the electron's charge, B is the noise-equivalent electrical bandwidth of the receiver, $\Delta\upsilon = 3.75\,$THz is the optical source bandwidth in Hertz, K_b is the Boltzmann's constant, $T_n = 300K$ is the absolute receiver noise temperature, and $R_L = 1030\,\Omega$ is the receiver load resistor.

Using Gaussian approximation, BER can be expressed as:

$$BER = P_e = \tfrac{1}{2}\,erfc\left(\sqrt{\tfrac{SNR}{8}}\right) \tag{9}$$

The system performance is shown in figure (6) for different MQC code size for two data rates. Data rate of 155 Mb/s shows good performance compared to 622 Mb/s. In communication systems, there is a trade-off between data bit rate and the provided system number of channels. Data bit rate x sequence code length = encoded chip rate. Generally, in optical CDMA analysis, in order to reduce the MAI limitations the data bit rate should be reduced. Increasing the bit rate will decrease the required average SNRs to maintain low BERs values, making the signal to be more sensitive to fiber dispersion and receiver circuitry noise.

The per code chip eavesdropper's SNRs as a function of the theoretical system capacity are shown in figure (7). If the authorized users transmit sufficient power so that 50%, 75%, 82%, and 85% of the theoretical system capacity is attained for MQC codes that have prime number p of 3, 7, 11, and 13 respectively, the eavesdropper has SNR of 15 dB. An optical matched filter receiver followed by envelope detection theoretically requires a peak SNR of approximately 15 dB to produce the required raw detector BER of 10^{-4}. Error correction codes used in commercial high-rate optical telecommunication equipment can produce the maximum acceptable system BER 10^{-9}.

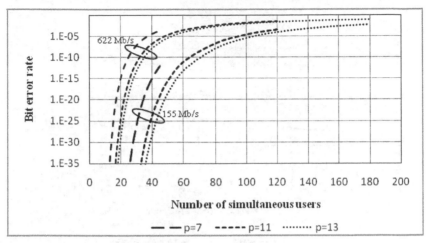

Fig. 6. BER versus number of simultaneous users. $P_{sr} = -10$ dBm.

The figure above shows a contradiction between network system performance and security. Increasing the network system capacity will lead the eavesdropper to detect high SNRs. Another limitation can be shown in figure (8), where high specified SNRs will increase the eavesdropper possibility of attacks.

Thus, for secure firms, a network designer should take these limitations under consideration. If 50% of the system capacity is provided, specified authorized SNRs between 10 dB to 15 dB are suitable for eavesdropper to get encoded pulse SNRs between 10 dB and 15 dB, respectively. Their corresponding bit error rates BERs are nearly 10^{-5} and 10^{-2}, respectively as shown in figure (9).

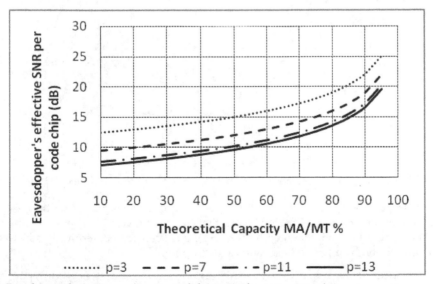

Fig. 7. Per chip code SNR as a function of theoretical system capacity

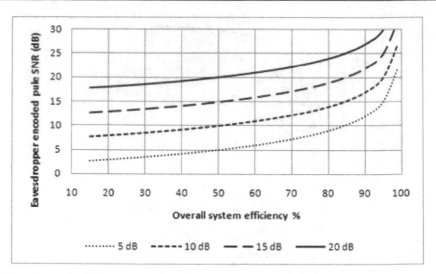

Fig. 8. Per chip code SNR as a function of theoretical system capacity for different specified authorized SNRs

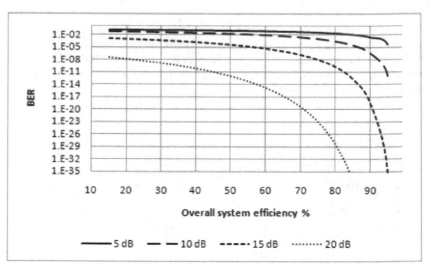

Fig. 9. BERs as a function of theoretical system capacity for different specified authorized SNRs

The eavesdropper performance of detecting spectral encoding chip bandwidth pulses form spectral amplitude optical CDMA code word that has been investigated in (Bakarman et al. 2009). The basic MQC code denoted by (12, 4, 1), has been considered to demonstrate the performance for both authorized user and eavesdropper.

Wide bandwidth enhances SNRs for both authorized user and eavesdropper, which increases the possibility of eavesdropping. Therefore, from the security viewpoint, one should minimize the eavesdropper ability to detect code word pulses by controlling the authorized performance to reasonable throughput. This leads to security impact over system performance as shown in figure (10). The solid and dashed lines represent theoretical results for authorized user and eavesdropper, respectively using MQC (12, 4, 1). Whereas, triangle and rectangle symbols represent results for authorized user and eavesdropper, respectively using M. sequence code (7, 4, 2).

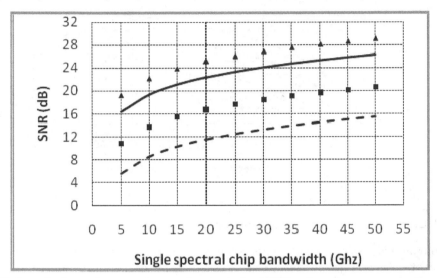

Fig. 10. Security impact over system performance for MQC code system

Thus, to improve the degree of security, we have to reduce the bandwidth of the encoding chip bandwidth pulses. This reduction should not affect the system performance. For example, if a spectral chip is reduced from 50 GHz to 25 GHz, the authorized user and eavesdropper could obtain SNRs of 23 dB and 12 dB respectively. These values correspond to bit error rate BERs of nearly 10^{-12} and 10^{-4} respectively. The maximum acceptable system BER is assumed to be 10^{-9}. Decreasing spectral chip, below than 25 GHz, will affect the authorized user performance forcing him to use error correction codes techniques used in commercial optical communications.

The results show that using unipolar optical CDMA codes schemes based on MQC and modified double weight MDW (Aljunid et al. 2004) code system enhance the security with a low cost implementation in comparison to the bipolar ones based on modified pseudorandom noise (PN) code (Chung et al. 2008), see also figure (10). MQC (12, 4, 1) code has 5 dB security preferences over PN (7, 4, 2) code. For the authorized users, bipolar codes would show high performance in comparison to unipolar codes because the bipolar signaling has a 3-dB signal-to-noise ratio (SNR) advantage over the on-off keying system with high cost implementation because each transmitter sends energy for both "0" and "1" bit (Nguyen et al. 1995). From the security viewpoint, one should minimize the eavesdropper ability to detect code word pulses by controlling the authorized performance to reasonable throughput.

Further security enhancement can be obtained by increasing the code dimension as shown in figure (11). With large value of prime number p, the main parameter to construct MQC codes, the eavesdropper ability to detect single encoded pulses becomes difficult even with wideband spectral chip. The eavesdropper BER will be higher than 10^{-3}.

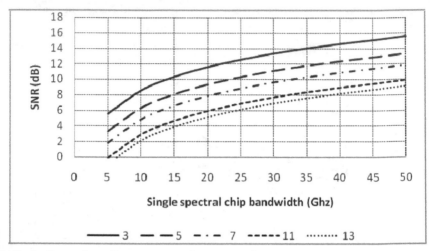

Fig. 11. Code dimension effects on eavesdropper performance

In communication systems, there is a tradeoff between data bit rate and the provided system number of channels. Data bit rate x sequence code length = encoded chip rate. Generally, in optical CDMA analysis, in order to reduce the MAI limitations, the data bit rate should be reduced. Figure (12) shows the impact of data bit rates on the eavesdropper performance. Increasing the bit rate will decrease the eavesdropper SNR, making the signal to be more sensitive to fiber dispersion and receiver circuitry noise.

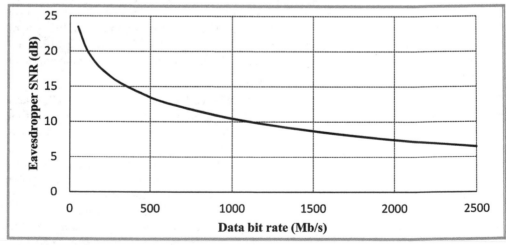

Fig. 12. Eavesdropper SNR vs bit rates

5. Conclusion

Improving the degree of security or enhancing the performance of optical CDMA networks have their impacts on each other such that a decision has to be made for cases in which all or some of these requirements are desired but cannot be fulfilled. The tradeoffs between security and the performance in optical CDMA, based on Modified Double Weight (MQC) system, are presented. From the security viewpoint, optical CDMA designer should minimize the eavesdropper ability to detect code word pulses by controlling the authorized performance to reasonable throughput. Otherwise, error correction codes techniques used in commercial optical communications would be the solution to obtain the maximum acceptable system BER.

6. Acknowledgment

This work was carried out at the Photonics Technology Laboratory (PTL), Institute of Micro Engineering and Nanoelectronics (IMEN),Universiti Kebangsaan Malaysia (UKM), under the supervision of professor Sahbudin Shaari. I would like to express my gratitude to him for providing a conductive enviroment for performing this research at this institute.

7. References

Aljunid, S. A., Ismail, M., Ramli, A. R., Ali, B. M. & Abdullah, M. K. 2004. A new family of optical code sequences for spectral-amplitude-coding optical CDMA systems. *Photonics Technology Letters, IEEE* 16 (10): 2383-2385.

Bakarman, H. A., Shaari, S. & Ismail, M. 2009. Security Performance of Spectral Amplitude Code OCDMA: Spectrally Encoded Pulse Bandwidth Effects. *J. Opt. Commun.* 30 (4): 242-247

Castro, J. M., Djordjevic, I. B. & Geraghty, D. F. 2006. Novel super structured Bragg gratings for optical encryption. *Lightwave Technology, Journal of* 24 (4): 1875-1885.

Chung, H. S., Chang, S. H., Kim, B. K. & Kim, K. 2008. Experimental demonstration of security-improved OCDMA scheme based on incoherent broadband light source and bipolar coding. *Optical Fiber Technology* 14 (2): 130-133.

Fisch, E. A. & White, G. B. 2000. *Secure Computers and Networks: Analysis, Design, and Implementation*. Boca Raton, FL: CRC Press LLC.

Imai, H., Rahman, M. G. & Kobara, K. 2005. *Wireless Communications Security*: Artech House Universal Personal Communications.

Jin-Hee, C. & Ing-Ray, C. On design tradeoffs between security and performance in wireless group communicating systems, at. *Secure Network Protocols, 2005. (NPSec). 1st IEEE ICNP Workshop on*: 13-18. 6 Nov. 2005

Mahafza, B. R. & Elsherbeni, A. 2003 *MATLAB Simulations for Radar Systems Design* Boca Raton: CHAPMAN & HALL/CRC.

Nguyen, L., Aazhang, B. & Young, J. F. 1995. All-optical CDMA with bipolar codes. *Electronics Letters* 31 (6): 469-470.

Oyster Optics, I. 2008. *http://www.oysteroptics.com/index_resources.html*.

Peterson, R. L., Ziemer, R. E. & Borth, D. F. 1995. *Introduction to Spread Spectrum Communications*. Englewood Cliffs: Prentice Hall.

Salehi, J. A. 1989. Code division multiple-access techniques in optical fiber networks. I. Fundamental principles. *Communications, IEEE Transactions on* 37 (8): 824-833.

Salehi, J. A. & Brackett, C. A. 1989. Code division multiple-access techniques in optical fiber networks. II. Systems performance analysis. *Communications, IEEE Transactions on* 37 (8): 834-842.

Shake, T. H. 2005a. Confidentiality performance of spectral-phase-encoded optical CDMA. *Lightwave Technology, Journal of* 23 (4): 1652-1663.

Shake, T. H. 2005b. Security performance of optical CDMA Against eavesdropping. *Lightwave Technology, Journal of* 23 (2): 655-670.

Smith, E. D. J., Blaikie, R. J. & Taylor, D. P. 1998. Performance enhancement of spectral-amplitude-coding optical CDMA using pulse-position modulation. *Communications, IEEE Transactions on* 46 (9): 1176-1185.

Teixeira, A., Vieira, A., Andrade, J., Quinta, A., Lima, M., Nogueira, R., Andre, P. & Tosi Beleffi, G. Security issues in optical networks physical layer, at. *Transparent Optical Networks, 2008. ICTON 2008. 10th Anniversary International Conference on*: 123-126. 22-26 June 2008

Wolter, K. & Reinecke, P. 2010. Performance and Security Tradeoff. In. Aldini, A., Bernardo, M., Di Pierro, A. & Wiklicky, H. (eds.). *Formal Methods for Quantitative Aspects of Programming Languages*: 135-167 Springer Berlin / Heidelberg.

Zorkadis, V. 1994. Security versus performance requirements in data communication systems. In. Gollmann, D. (eds.). *Computer Security – ESORICS 94*: 19-30 Springer Berlin / Heidelberg.

Zou, W., Ghafouri-Shiraz, H. & Shalaby, H. M. H. 2001. New code families for fiber-Bragg-grating-based spectral-amplitude-coding optical CDMA systems. *Photonics Technology Letters, IEEE* 13 (8): 890-892.

Zou, W., Shalaby, H. M. H. & Ghafouri-Shiraz, H. 2001. Modified quadratic congruence codes for fiber Bragg-grating-based spectral-amplitude-coding optical CDMA systems. *Lightwave Technology, Journal of* 19 (9): 1274-1281.

7

Coherent Multilook Radar Detection for Targets in KK-Distributed Clutter

Graham V. Weinberg
Electronic Warfare and Radar Division
Defence Science and Technology Organisation (DSTO)
Australia

1. Introduction

1.1 The problem

This Chapter examines the problem of coherent multilook radar detection of targets in the sea from an airborne maritime surveillance platform. The radar of interest operates at high resolution, and at a high grazing angle, with frequency in X-Band, and is fully polarised. Typically, such surveillance radars are searching for small targets, such as fishing boats or submarine periscopes. When a radar transmits signals toward a potential target, the returned signal is distorted by a number of sources. The signal will be affected by thermal noise in the radar. Atmospheric conditions will also distort the signal. In a maritime surveillance context, backscatter from the rough sea surface will result in what is known as clutter returns Levanon (1988). In most cases, the first two sources of interference can be modelled by Gaussian processes Gini et al (1998). However, the sea clutter distribution is much more complex to model. Much research has consequently been dedicated to modelling sea clutter returns. Once a suitable clutter model has been proposed and validated, detection schemes can be designed using statistical methodology. This Chapter will construct radar detection schemes for a modern clutter model used for the scenario of interest. Before introducing this, a synopsis of clutter models is provided. Useful references on radar include Levanon (1988); Mahafza (1998); Peebles (1998); Stimson (1998). For statistical hypothesis testing, consult Beaumont (1980). For a detailed description of statistical and probability methods employed here, consult Durrett (1996). A useful reference on statistical distributions is Evans et al (2000).

1.2 Historical perspective

In earlier low resolution radar systems, sea clutter returns were found to be well modelled by a Rayleigh distribution for amplitude, with the exception of the case of low grazing angle scanning Shnidman (1999). As technology improved, and allowed the development of higher resolution radars, it was found that the Rayleigh assumption was no longer valid. As far back as 1967 it was found that data taken from a high resolution radar, operating at X-band with vertical polarization and a 0.002 • second pulse, and at a grazing angle of 4.7°, deviated from the Rayleigh assumption significantly Trunk and George (1970). The major reason for the failure in the Rayleigh assumption is that the higher the radar's resolution, the smaller the resolution cells, and the corresponding clutter densities have greater tails than the zero mean Gaussian clutter models investigated by Marcum and Swerling Trunk and George (1970).

Within small resolution cells the clutter has been observed to consist of a series of discrete spikes, or high intensity returns, that vary in time. As is also reported in Trunk and George (1970), failure to alter the clutter model for high resolution radar results in an increased incidence of false alarms, which can seriously undermine a radar's performance. Hence it became critical to develop new models for high resolution radar sea clutter.

Sea clutter return statistics have also been found to have a complex dependence on a given radar's operating characteristics. In particular, it has been found that such statistics depend on the radar's operating mode, grazing angle and background operating environment. The clutter returns can vary significantly with grazing angle: the critical grazing angle $\bullet_c = \frac{\bullet}{4 \bullet \bullet_h}$ divides the set of grazing angles into a "high" and "low" class. Here $\bullet$ is the wavelength and $\bullet_h$ is the root mean squared surface height deviation above average height Skolnik (2008). Clutter returns vary significantly over the partition of grazing angles by this threshold. At a low grazing angle, sea clutter returns are subjected to shadowing, ducting and multipath propogation Dong (2006); Skolnik (2008). By contrast, at higher grazing angles, the returns are affected by Bragg scattering from rough surface and scattering from whitecaps Dong (2006). It was reported in Crombie (1955) that at high frequency, wavelength scattering seemed to arise due to a resonant interaction with sea waves with one half of the incident wavelength. Such phenomenon is referred to as Bragg scattering. Whitecaps refer to the visible appearance of whitecaps in the sea surface as waves break Ward et al (2006). Wind speed affects the sea state, which in turn has a significant effect on clutter Skolnik (2008). The radar's polarization also has been found to affect the behaviour of clutter returns, with increased spikiness with horizontal polarization Dong (2006). Thus it is clear that adequate models for sea clutter backscattering must account for such phenomena as described above, and need to model the complex nature of true sea clutter returns.

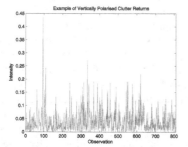 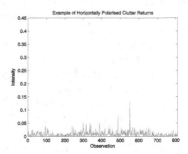

Fig. 1. Examples of real high grazing angle clutter. The left plot shows vertically polarized clutter, while the right is for horizontally polarization. All measurements are in intensity.

Figure 1 contains examples of real sea clutter returns, in the intensity domain, showing the changes in clutter as the radar's polarization is switched from vertical to horizontal, on a pulse to pulse basis. Both these examples of clutter have been obtained during a trial conducted by DSTO in the Southern Ocean using Ingara, an X-band fully polarized airborne radar Stacy et al (2003). The grazing angle for the clutter shown was $38.7°$. This has been taken from a range profile in data set run3468 at an azimuth angle of $225°$. Further details can be found in Stacy et al (2003), and will be provided with the numerical analysis to follow.

Earlier solutions to the inadequacy of the Rayleigh model include the Log-Normal Distribution, which was used for low grazing angles ($\bullet \leq 5°$)

Trunk and George (1970), and the Weibull Distribution Schleher (1976). As reported in the latter, the Weibull model was found to be a close fit to high resolution radar sea clutter returns with a grazing angle $1° \leq \bullet \leq 30°$. An attractive feature of the Weibull model is that it encompasses the two previous classical models of Rayleigh and Log-Normal.

However, it was found that the Weibull model was still insufficient because it did not take into account the complex nature of clutter. In fact, this model had been developed on the basis of empirical studies of clutter and not including any real physical understanding of the nature of sea clutter. Further investigation of high resolution sea clutter showed that time plots for an individual range cell demonstrated a fast pulse to pulse fluctuation. There was also an underlying slow modulation of the fast fluctuations. Presence of fast fluctuations and decorrelation with frequency agility implies there are many scatterers contributing to the resultant echo within each illumination patch. Hence an appropriate statistical model of clutter was needed to accommodate this phenomenon.

1.3 The K-Distribution

The K-Distribution was introduced in an attempt to model the phenomena described in the previous subsection. This model has a physical interpretation as well as a theoretical justification. This distribution models the observed fast fluctuations of clutter using a conditional Rayleigh distribution, while the underlying modulation is modelled through a Gamma distribution. Such a process can be described as a modulated Gaussian process. The short term or fast fluctuation is referred to as the speckle component, while the modulation or slow varying component is called the texture. The modern formulation of the K-Distribution can be traced back to Ward (1981), who arrived at it by considering high resolution sea clutter returns as a product of two independent random variables. One component has a short correlation time and is decorrelated by frequency agility (Rayleigh component referred to above). The other component has a long correlation time and models the underlying modulation or sea swell component (the Gamma distributed component).

The first appearance of the K-Distribution, applied to modelling radar clutter returns, is Jakeman and Pusey (1976). In the latter a mathematical model was developed for non-Rayleigh clutter, under the assumption that the illuminated area of the sea surface can be seen as a finite number of discrete scatterers. Each scatterer is assumed to give randomly phased contributions of the fluctuating amplitude in the radar's far field. The K-Distribution allowed explicit dependence on the illuminated area. In addition to this theoretical development, Jakeman and Pusey (1977) validated the K-Distribution as a model for X-band sea clutter taken from a coastal cliff site in England.

We now introduce the K-Distribution mathematically. The reference Ward et al (2006) is particularly useful in deriving K-Distribution properties.

The K-Distribution $X = [X_1, X_2]^\mathsf{T}$ is defined, in the complex domain, as a bivariate random variable through the product

$$X = \sqrt{\frac{2\bullet}{\bullet}} X_{IQ}, \tag{1}$$

where $\bullet = \bullet(\bullet, \bullet)$ is a Gamma random variable with density

$$f_\bullet(t) = \frac{1}{\Gamma(\bullet)} \left(\frac{\bullet}{\bullet}\right)^\bullet t^{\bullet-1} e^{-\frac{\bullet t}{\bullet}}, \tag{2}$$

for $t \geq 0$, and X_{IQ} is a bivariate Gaussian random variable, with mean the 2×1 zero vector 0, covariance matrix the 2×2 identity matrix I_2 and density

$$f_{X_{IQ}}(x) = \frac{1}{2\bullet} e^{-\frac{1}{2}x^T x}, \tag{3}$$

where x^T is the transpose of x. The two components of X_{IQ} represent the in-phase and quadrature elements of the complex clutter return sampled by the radar. The Gamma distribution (2) has two free variables $\bullet$ and $\bullet$. The former variable is an intermediate scale parameter, while $\bullet$ is called a shape parameter. The $\sqrt{\bullet}$ component in (1) is the slow varying component, while the Gaussian element is the fast varying component.

The representation (1) is convenient mathematically, from a signal processing point of view, when considering targets embedded in K-Distributed clutter and Gaussian noise. This is because, under the assumption of independent clutter and noise components in the return, we can use statistical conditioning on $\bullet$ to combine these into one Gaussian component. This facilitates the analysis considerably: see Conte et al (1995); Farina et al (1995) for examples of this.

The amplitude of X is

$$K := |X| = \sqrt{\frac{2\bullet}{\bullet}} |X_{IQ}| = \sqrt{\frac{2\bullet}{\bullet}} R, \tag{4}$$

where the norm $|\cdot|$ in (4) is the standard Euclidean norm on vectors and $R := |X_{IQ}|$.

The density of (4) can be derived by applying statistical conditioning; see Durrett (1996). In particular, by conditioning on the variable $\bullet$,

$$f_K(t) = \int_0^\infty f_{K|\bullet}(t|\bullet) f_\bullet(\bullet) d\bullet. \tag{5}$$

With a considerable amount of analysis, it can be shown that

$$f_K(t) = \frac{2c}{\Gamma(\bullet)} \left(\frac{ct}{2}\right)^\bullet K_{\bullet-1}(ct), \tag{6}$$

where $c = \sqrt{\frac{\bullet\bullet}{\bullet}}$, giving the density function of the K-Distribution's amplitude. The parameter $\bullet$, which arises from the underlying Gamma Distribution, is referred to as the K-Distribution's shape parameter, while c is called the scale parameter. The shape parameter $\bullet$ governs the tail of the K-Distribution's density, and it has been found that small values of $\bullet$ represent more spiky clutter Crisp et al (2009). Larger values of $\bullet$ produce backscattering that is closer to Rayleigh in distribution. As reported in Crisp et al (2009), $\bullet$ around 0.1 corresponds to very spiky clutter, while $\bullet$ near 20 produces approximate Rayleigh clutter.

The cumulative distribution function of the K-Distribution can be obtained by integrating (6) over the interval $[0, t]$. It can be shown that

$$F_K(t) = P(K \leq t) = 1 - \frac{(ct)^\bullet K_\bullet(ct)}{2^{\bullet-1}\Gamma(\bullet)}. \tag{7}$$

As reported in Dong (2006), the first two moments of K are given by

$$E(K) = \frac{\sqrt{\bullet}\,\Gamma(\bullet + \frac{1}{2})}{c\Gamma(\bullet)} \quad \text{and} \quad E(K^2) = \frac{4\bullet}{c^2}, \tag{8}$$

and consequently the K-Distribution's variance can be obtained as

$$\text{Var}(K) = E(K^2) - [E(K)]^2 = \frac{1}{c^2}\left[4\bullet - \frac{\bullet\,\Gamma^2(\bullet + 0.5)}{\Gamma^2(\bullet)}\right]. \tag{9}$$

1.4 Modifying the K-Distribution: The KA-Distribution

As remarked in the previous subsections, the K-Distribution was introduced in an attempt to reflect the complex nature of sea clutter backscattering of high resolution radars. However, it has been found that horizontally polarised data tends to be spikier than that obtained from a vertically polarized radar. In the latter, the K-Distribution is a suitable model, but in the former, there is a deviation in the distribution's tail region. Hence, this resulted in the search for a new model that incorporated such deviations in the distribution's fit.

A major step toward this objective was the introduction of the KA-Distribution in Middleton (1999). In this model, the fast varying component is still Rayleigh Distributed, but it is conditioned on the number of spikes in each range cell. The latter is assumed to follow a Poisson Distribution. The slow varying component is still Gamma Distributed, which is the local intensity of Bragg/whitecap scatterers. Mathematically, the KA-Distribution is defined in an identical formulation to (5). Specifically, its density is given by the integral

$$f_{KA}(x) = \int_0^\infty f_{KA|\bullet}(x|t)\,f\bullet(t)dt, \tag{10}$$

where $\bullet$ has a Gamma Distribution with parameters $\bullet$ and $\bullet = \bullet_{bw}$ (refer to (2)). The conditional distribution $KA|\bullet$ has density

$$f_{KA|\bullet}(x|t) = \sum_{n=0}^\infty \frac{2x}{t + \bullet n + n\bullet_{sp}} e^{-\left(\frac{x^2}{t+\bullet n+n\bullet sp}\right)} \bullet(n), \tag{11}$$

where $\bullet(n) = \frac{e^{-\bar{N}}\bar{N}^n}{n!}$ is the Poisson probability that a range cell contains n spikes. In (11), $\bullet_n$ is the mean noise intensity, $\bullet_{sp}$ is the spike intensity, $\bullet_{bw}$ is the mean Bragg/whitecap intensity and $\bar{N}$ is the mean number of spikes in each range cell Dong (2006). As also pointed out in the latter, the noise, spikes and Bragg/whitecap scatterers are all assumed to be mutually uncorrelated. The expression (11) is the law of total probability applied over all possible number of spikes in range cells.

The KA-Distribution was also investigated in Ward and Tough (2002) and Watts et al (2005), and was shown to improve the fit of sea clutter in the distribution's tail region. Further validation of the KA-Distribution's fit to spiky sea clutter is given in Valeyrie et al (2009) using airborne radar records in rough sea states. Unfortunately, there is no closed form expression for the density of the KA-Distribution.

1.5 The KK-Distribution

The KK-Distribution was proposed in Dong (2006) as an alternative to the KA-Distribution. This Distribution assumes both the Bragg/whitecap scatterers and spikes are K-Distributed. The overall clutter distribution is defined as a mixture of the two: namely, a weighted sum of densities of the respective components Dong (2006). This formuation leads to a closed form for the density, in the amplitude domain, in terms of K-Distribution densities, and statistical properties can be determined via linearity of the density, as will be demonstrated below.

In particular, for a fixed parameter $k \in [0,1]$, the amplitude density of the KK-Distribution is defined as the mixture

$$f_{KK}(t) = (1-k)f_{K_1}(t; \bullet, \bullet) + kf_{K_2}(t; \bullet_{sp}, \bullet_{sp}), \tag{12}$$

where each f_{K_j} is a K-Distribution with parameters $(\bullet, \bullet)$ and $(\bullet_{sp}, \bullet_{sp})$ respectively. The first K-Distribution density in (12) represents the Bragg and whitecap scatterers in the model. The second K-Distribution in (12) represents the spike component of the clutter.

The restricted range of values for k ensures (12) preserves the features of a probability density. As reported in Dong (2006), it has been found empirically that we may assume $\bullet = \bullet_{sp}$, and so selection of k, $\bullet_{sp}$ and $\bullet := \frac{\bullet_{sp}}{\bullet}$ determines the spike component. Throughout we will use the notation $c_1 = \bullet$ and $c_2 = \bullet_{sp}$ for the two scale parameters.

Comparison of this model to high resolution radar sea clutter has been recorded in Dong (2006), where it is shown the KK-Distribution provides a better fit to the upper tail region of the empirical distribution than the K- and KA-Distributions. The KK-Distribution's validity is also supported by the analysis of trials data in Rosenberg et al (2010), who also extend the model to include multiple looks and thermal noise.

The moments of the KK-Distribution can be obtained from those of the K; in particular, if we let KK be a random variable with the density (12), it follows that

$$E(KK) = (1-k)E(K_1) + kE(K_2) = \frac{\sqrt{\bullet}\,\Gamma(\bullet)}{\Gamma(\bullet + \frac{1}{2})}\left(\frac{1-k}{c_1} + \frac{k}{c_2}\right), \tag{13}$$

where we have applied (8) for two K-Distributions with common shape parameter $\bullet$ and scale parameters c_1 and c_2 respectively. The KK-Distribution's second moment can also be derived from (8): in particular, and under the same parameter values, we can show

$$E(KK^2) = (1-k)E(K_1^2) + kE(K_2^2) = 4\bullet\left(\frac{1-k}{c_1^2} + \frac{k}{c_2^2}\right). \tag{14}$$

We can use (13) and (14) to derive the KK-Distribution's variance. It is straightforward to derive

$$\text{Var}(KK) = 4\bullet\left(\frac{1-k}{c_1^2} + \frac{k}{c_2^2}\right) - \frac{\bullet\,\Gamma^2(\bullet)}{\Gamma^2(\bullet + \frac{1}{2})}\left(\frac{1-k}{c_1} + \frac{k}{c_2}\right)^2. \tag{15}$$

The cumulative probability density function of the KK-Distribution can be obtained from the fact that

$$F_{KK}(t) = (1-k)F_{K_1}(t) + kF_{K_2}(t)$$

$$= 1 - \frac{t^\bullet}{2^{\bullet-1}\Gamma(\bullet)}\left((1-k)K_\bullet(c_1 t)c_1^\bullet + kK_\bullet(c_2 t)c_2^\bullet\right), \tag{16}$$

where the parameter sets $\{c_1, \bullet\}$ and $\{c_2, \bullet\}$ are for the two K-Distributions K_1 and K_2 respectively.

1.6 Scope, assumptions and structure

This Chapter will be focused on multilook radar detection of targets embedded within KK-Distributed clutter. In order to do this, a number of assumptions will be made. Throughout, the thermal noise and Gaussian interference will be ignored. It will be assumed that the radar has the capability to determine the clutter parameters, so that these are known from the detection point of view. The clutter covariance matrix will also be assumed known, and furthermore, it will be assumed its inverse is semi-positive definite. It will be assumed that the normalised Doppler frequency (to be specified later) is also known. The Neyman-Pearson approach will be employed to construct detection schemes. The focus will be restricted to coherent radar detection.

Section 2 formulates the KK-Distribution as an intensity distribution of a complex spherically invariant random process. This enables the determination of the Neyman-Pearson optimal decision rule in Section 3. The generalised likelihood ratio test will be used to construct a suitable suboptimal decision rule, which assumes the target is constant on a scan to scan basis. Section 4 introduces the Ingara data briefly, and investigates receiver operating characteristics curves. These provide a means of gauging the performance of a detction scheme over varying signal to clutter strengths.

Throughout we will use P to denote probability, E to be expectation with respect to P, and $X \stackrel{d}{=} Y$ will represent equivalence in distribution.

2. Coherent multilook detection

2.1 Spherically invariant random processes

Spherically Invariant Random Processes (SIRPs) Rangaswamy et al (1993); Wise (1978); Yao (1973) provide a general formulation of the joint density of a non-Gaussian random process, enabling the construction of densities for Neyman-Pearson detectors Beaumont (1980); Neyman and Pearson (1933).

Traditionally, a SIRP is introduced as a process whose finite order subprocesses, called Spherically Invariant Random Vectors (SIRVs), possess a specific density Conte and Longo (1987); Rangaswamy et al (1991). However, due to an equivalent formulation, we can specify SIRVs and SIRPs in a manner more intuitive to the modelling of radar returns as follows. Let $c = \{\bullet_1, \bullet_2, \ldots, \bullet_N\}$ be the complex envelope of the clutter returns. Then this vector is called SIRV if it can be written in the compound-Gaussian formulation

$$c = S\mathcal{G}, \tag{17}$$

where the process $\mathcal{G} = (G_1, G_2, \ldots, G_N)^\mathsf{T}$ is a zero mean complex Gaussian random vector, or multidimensional complex Gaussian process, and S is a nonnegative real valued univariate random variable with density f_S. The latter random variable is assumed to be independent of the former process. If such a decomposition exists for every $N \in \mathbb{N}$ the complex stochastic process $\{\bullet_1, \bullet_2, \ldots\}$ is called spherically invariant. For a comprehensive description of the modelling of clutter via SIRPs consult Conte and Longo (1987). What is clear from the formulation (17) is that, by conditioning on the random variable S, we can write down the density of c as a convolution. This will be shown explicitly in the analysis to follow.

2.2 General formulation

The detection decision problem is now formulated. We assume that we have a SIRP model for the clutter c as specified through the product formulation (17). Suppose the radar return is z, which is a complex $N \times 1$ vector. Then the coherent multilook detection problem can be cast in the form

$$H_0 : z = c \text{ against } H_1 : z = Rp + c, \tag{18}$$

where all complex vectors are $N \times 1$, and H_0 is the null hypothesis (return is just clutter) and H_1 is the alternative hypothesis (return is a mixture of signal and clutter). Statistical hypothesis testing is outlined in Beaumont (1980). Here, the vector p is the Doppler steering vector, whose components are given by $p(j) = e^{j2\pi f_D T_s}$, for $j \in \{1, 2, \dots, N\}$, where f_D is the target Doppler frequency and T_s is the radar pulse repetition interval. It will be assumed that this is completely known. The complex random variable R accounts for target characteristics, and $|R|$ is the target amplitude. Suppose the zero mean Gaussian process has covariance matrix $\text{Cov}(\mathcal{G}) = E(\mathcal{G}\mathcal{G}^H) = \Sigma$. Since it is a covariance matrix, its inverse will exist. Recall that (since Σ^{-1} is also a covariance matrix), the inverse will be symmetric and positive definite. If we assume it is semi-definite positive, then we can apply a whitening filter approach to simplify the detection problem. Hence we suppose the Cholesky Factorisation exists for Σ^{-1}, so that there exists a matrix A such that $\Sigma^{-1} = A^H A$.

A SIRP is unaffected by a linear transformation Rangaswamy et al (1993). This means that applying a linear operator to the clutter process alters the complex Gaussian component, but the characteristic function of the SIRP is preserved. Hence in the literature, a whitening approach is often applied to the detection problem of interest. Note that, by applying the Cholesky factor matrix A to the statistical test, we can reformulate (18) in the statistically equivalent form

$$H_0 : r = n \text{ against } H_1 : r = Ru + n, \tag{19}$$

where $r = Az, n = Ac$ and $u = Ap$.

The transformed clutter process $n = SA\mathcal{G}$, and $A\mathcal{G}$ is still a multidimensional complex Gaussian process, with zero mean but covariance $\text{Cov}(A\mathcal{G}) = E(A\mathcal{G}\mathcal{G}^H A^H) = A\Sigma A^H$.

Since $\Sigma^{-1} = A^H A$, it follows that $I_{N \times N} = \Sigma A^H A$, from which it is not difficult to deduce that $A = (A\Sigma A^H)A$. Let $B = A\Sigma A^H$, then note that $B^2 = B$, so that B is idempotent. Also, it follows that B must be invertible, since $\det(B) = \det(A)\det(\Sigma)\det(A^H)$ and $\det(\Sigma) \neq 0$ and $\det(\Sigma^{-1}) = \det(A^H)\det(A) \neq 0$. Thus it follows that B must be the $N \times N$ identity matrix. Consequently, we conclude that the transformed Gaussian process $A\mathcal{G}$ has zero mean and covariance matrix this identity matrix. Hence the clutter, conditioned on the variable S, is completely decorrelated through this linear transform.

We write this as $A\mathcal{G} \stackrel{d}{=} CN(\vec{0}, I_{N \times N})$. Observe that $n|S$ is still complex Gaussian with zero mean vector but covariance matrix $s^2 I_{N \times N}$, hence its density is

$$f_{n|S=s}(x) = \frac{1}{\pi^N s^{2N}} e^{-s^{-2}\|x\|^2}. \tag{20}$$

Hence by using conditional probability Durrett (1996)

$$f_{\mathbf{n}}(\mathbf{x}) = \int_0^\infty f_{\mathbf{n}|S=s}(\mathbf{x}) f_S(s) ds$$

$$= \int_0^\infty \frac{1}{\bullet^N s^{2N}} e^{-s^{-2}\|\mathbf{x}\|^2} f_S(s) ds. \qquad (21)$$

Hence if we define a function $h_N(p)$ by

$$h_N(p) = \int_0^\infty s^{-2N} e^{-s^{-2}p} f_S(s) ds, \qquad (22)$$

then the clutter joint density can be written in the compact form

$$f_{\mathbf{n}}(\mathbf{x}) = \frac{1}{\bullet^N} h_N(\|\mathbf{x}\|^2). \qquad (23)$$

The function h_N defined in (22) is of paramount importance in SIRP theory, as all densities of interest are expressed in terms of it. It is thus called the characteristic function. Much of the literature is devoted to determining h_N and f_S pairs for particular desired clutter models Rangaswamy et al (1991; 1993).

Next we derive the marginal amplitude and intensity distributions of the complex clutter process $\mathbf{n}$, which are intimately related to the special function h_N. Let the kth element of $\mathbf{n}$ be n_k, so that $n_k = SA\mathcal{G}_k$, where $\mathcal{G}_k$ is the kth component of $\mathcal{G}$, for $1 \le k \le N$. Then the complex Gaussian density of $A\mathcal{G}_k$ is

$$f_{A\mathcal{G}_k}(z) = \frac{1}{\bullet} e^{-|z|^2}. \qquad (24)$$

Hence, equivalently, $A\mathcal{G}_k$ is a bivariate Gaussian process with zero mean and covariance matrix $\frac{1}{2} I_{2\times2}$. Consequently, it follows that its amplitude $|A\mathcal{G}_k|$ has a Rayleigh distribution with parameter $\frac{1}{\sqrt{2}}$. Hence, using the fact that S and $A\mathcal{G}_k$ are independent, and the definition of the Rayleigh density, we can show that

$$f_{|n_k|}(t) = \int_0^\infty \frac{2t}{s^2} e^{-\frac{t^2}{s^2}} f_S(s) ds = 2t h_1(t^2), \qquad (25)$$

yielding the marginal amplitude distribution for each k. Transforming to the intensity domain, it is not difficult to show

$$f_{|n_k|^2}(t) = \frac{1}{2} t^{-\frac{1}{2}} f_{|n_k|}(\sqrt{t}) = h_1(t), \qquad (26)$$

also for each k.

The importance of the results (25) and (26) is that they indicate how an arbitrary distribution can be embedded within a SIRP model. The key to the determination of a relevant SIRP for a given detection problem is the specification of the random variable S through its density f_S. For a desired marginal distribution, we can form an integral equation using (26) and (22) with $N = 1$. The literature contains extensive discussion of this problem; in particular, Rangaswamy et al (1993) contains guidelines on performing this, as well as case studies for most desired marginal distributions.

2.3 The KK-SIRP

The key to constructing the KK-SIRP is to observe the fact that, since it is defined as a mixture in the amplitude domain, we can use the K-Distribution SIRP, which is well known. This is specified in Rangaswamy et al (1993). The following specifies the density of S for a K-Distribution:

Lemma 2.1. The SIRP with nonnegative random variable S with density given by

$$f_S(s) = \frac{\sqrt{2}c^{2\bullet}}{\Gamma(\bullet)} s^{2\bullet-1} e^{-\frac{c^2 s^2}{4}}, \tag{27}$$

admits a K-Distribution, with parameters c and •, as marginal amplitude distributions.

Extending this result to the KK-Distribution case is simple, as the following Lemma encapsulates:

Lemma 2.2. The function

$$f_S(s) = (1-k)f_{S_1}(s) + kf_{S_2}(s), \tag{28}$$

with $k \in [0, 1]$ and each $f_{S_i}(s)$ given by (27) with parameters c_i and $\bullet_i$, is a well-defined density on the nonnegative real line. Furthermore, it admits a KK-Distribution with scale parameters c_i and shape parameters $\bullet_i$.

The proof of Lemma 2.2 is not difficult, and is hence omitted. It requires the application of Lemma 2.1, together with the following result:

Lemma 2.3. For the SIRP generated by the density (28),

$$h_N(u) = \frac{u^{\frac{nu-N}{2}} \bullet (\sqrt{u}; c_1, c_2, \bullet, k)}{\Gamma(\bullet)2^{1.5\bullet+0.5N-1}}, \tag{29}$$

with • defined by

$$\bullet(u; c_1, c_2, \bullet, N, k) = (1-k)c_1^{N+\bullet} K_{N-\bullet}(c_1 u) + kc_2^{N+\bullet} K_{N-\bullet}(c_2 u). \tag{30}$$

The proof of Lemma 2.3 involves simplification of the characteristic function (22), application of Lemma 2.1 and use of the corresponding properties of the K-Distribution SIRP Rangaswamy et al (1993).

3. Detector decision rules

Using the KK-SIRP formulated in the previous section, we can now derive the Neyman-Pearson optimal decision rules, including suboptimal approximations.

3.1 Optimal decision rule: constant target

It is useful to begin with the case of a fixed known target model, which means we assume that R is a fixed constant that is completely specified. This enables one to specify the exact form of the Neyman-Pearson optimal detector. Throughout, we will also assume the clutter parameters have been estimated by the radar system, so that they are completely known. It is also assumed that the clutter parameters are homogeneous within a range profile scan.

Under H_1, $r = Ru + n$, and conditioned on S, this is still complex Gaussian, with the same covariance as n (since Ru is constant), but its mean is shifted by the vector Ru. Hence the density under H_1 is

$$f_{H_1}(r) = \frac{1}{\bullet^N} h_N(\|r - Ru\|^2). \tag{31}$$

The Neyman-Pearson optimal detector is the ratio of the densities under H_1 and H_0 Beaumont (1980). The density under H_0 is given by (23), with an application of (29). Hence, combining these results, we get the likelihood function

$$L(r) = \frac{h_N(|r - Ru|^2)}{h_N(|r|^2)}$$

$$= \left(\frac{|r - Ru|}{|r|}\right)^{\bullet - N} \frac{\bullet(|r - Ru|; c_1, c_2, \bullet, N, k)}{\bullet(|r|; c_1, c_2, \bullet, N, k)}, \tag{32}$$

and the optimal decision rule is

$$L(r) \underset{H_0}{\overset{H_1}{\gtrless}} \bullet, \tag{33}$$

where $\bullet$ is the detection threshold, which can be determined numerically from the false alarm probability. The notation $X \underset{H_0}{\overset{H_1}{\gtrless}} Y$ means that we reject H_0 if and only if $X > Y$. Hence for a radar return r we compare $L(r)$ to the threshold $\bullet$ to make a decision on whether a target signature is present in the return.

In all reasonable practical applications, we will not have knowledge of R, and so we must extend this result to account for an unknown R. This is the subject of the next Subsection.

3.2 Generalised likelihood ratio test

In a real application, the target is unknown and hence R must be estimated. If we assume R is unknown but constant from scan to scan, the Generalised Likelihood Ratio Test is used to first estimate this parameter, and then apply the estimate to the test (33). The methodology of GLRT is described in Barnard and Khan (2004); Conte et al (1995). We essentially produce a suboptimal detector for the case of an unknown target, based upon the optimal decision rule formulated for a completely known target model.

We call this estimate $\hat{R}$, which is taken to be the maximum likelihood estimator Beaumont (1980). It is chosen to minimise the quantity $\|r - Ru\|^2$. This is because, in view of the density (31) and the definition of h_N in (22), maximising the density with respect to R is equivalent to minimisation of this quantity.

By applying a simple expansion, we can show

$$\|r - Ru\|^2 = \|r\|^2 + |R|^2\|u\|^2 - 2\Re(\bar{R}u^H r), \tag{34}$$

where $\Re(z)$ is the real part of the complex number z. Recalling the proof of the Cauchy-Schwarz inequality (see Kreyszig (1978) for example), this is minimised with the choice $\hat{R} = \frac{u^H r}{\|u\|^2}$. Substituting this in (34) shows that

$$\min_R \|r - Ru\|^2 = \|r - \hat{R}u\|^2 = \|r\|^2 - \frac{|u^H r|^2}{\|u\|^2}. \tag{35}$$

An application of (35) to (33) results in the GLRT.

4. Numerical analysis of detectors

4.1 The Ingara data

Before proceeding with an analysis of the optimal and suboptimal detectors, the Ingara data is discussed briefly. This sea clutter set was collected by DSTO, using their radar testbed called Ingara. This is an X-Band fully polarised radar. The 2004 trial used to collect the data was located in the Southern Ocean, roughly 100km south of Port Lincoln in South Australia. Details of this trial, the Ingara radar, and data analysis of the sea clutter can be found in Crisp et al (2009); Dong (2006); Stacy et al (2003; 2005).

The radar operated in a circular spotlight mode, so that the same patch of sea surface was viewed at different azimuth angles. The radar used a centre frequency of 10.1 GHz, with 20 • s pulse width. Additionally, the radar operated at an altitude of 2314m for a nominal incidence angle of 50°, and at 1353m for 70° incidence angle. The trial collected data at incidence angles varying from 40° to 80°, on 8 different days over an 18 day period. As in Dong (2006), we focus on data from two particular flight test runs. These correspond to run34683 and run34690, which were collected on 16 August 2004 between 10:52am and 11:27am local time Dong (2006). Dataset run34683 was obtained at an incidence angle of 51.5°, while run34690 was at 67.2°. In terms of grazing angles, these correspond to 38.7° and 22.8° respectively. Each of these datasets were also processed in blocks to cover azimuth angle spans of 5° over the full 360° range. Roughly 900 pulses were used, and 1024 range compressed samples for each pulse were produced, at a range resolution of 0.75m. In Dong (2006) parameter estimates for the data sets run34683 and run34690 are given, enabling the fitting of the K- and KK-Distributions to this data. This will be employed in the numerical analysis to follow.

As reported in Dong (2006), the radar is facing upwind at approximately 227° azimuth, which is the point of strongest clutter. Downwind is at approximately 47°, which is the point where the clutter is weakest. Crosswind directions are encountered at 137° and 317° approximately.

4.2 Target model and SCR

Since the Ingara clutter does not contain a target, an artificial model is used to produce the receiver operating characteristics (ROC) curves. These curves plot the probability of detection against the signal to clutter ratio (SCR). It hence indicates the performance of a detector relative to varying signal strengths in the clutter model.

Throughout, a Swerling 1 target model is used Levanon (1988). This is equivalent to assuming that the in-phase and quadrature components of the signal return are complex Gaussian in distribution. For the problem under investigation, the SCR is given by

$$SCR = \frac{E(|R|^2)\,|u|^2}{E(S^2)N},$$
(36)

and it is not difficult to show that

$$E(S^2) = 4 \cdot \left[\frac{1-k}{c_1^2} + \frac{k}{c_2^2} \right].$$
(37)

Hence, assuming for the Swerling 1 target model, $E|R|^2 = 2\bullet^2$, an application of this to (36) yields

$$SCR = \frac{\bullet^2 |u|^2}{2N\bullet \left[\frac{1-k}{c_1^2} + \frac{k}{c_2^2}\right]}.$$
(38)

The ROC curves to follow are produced by determining the parameter $\bullet^2$, for a given SCR, using (38).

4.3 Receiver operating characteristics curves

Four examples of detector performance are provided. In all cases, the ROC curves have been produced using Monte Carlo simulations, with approximately 10^5 runs. Each simulation is for the case of a false alarm probability of 10^{-6}. The SCR is varied from -10 to 30 dB. Each ROC curve shows the performance of the optimal decision rule (33), the GLRT decision rule using (35) to estimate the target strength and the performance of the whitening matched filter. The latter is the optimal detector for targets in Gaussian distributed clutter, and can be used as a suboptimal decision rule. For a return r, it is given by

$$M(r) = \left|u^H r\right|^2,$$
(39)

where u^H is the Hermitian transpose. Although the decision rule (33) is dependent on the target parameters, and so is not useful in practice, it is used here to gauge the performance of the two suboptimal decision rules.

Each simulation uses parameters estimated from a specific Ingara data set. The clutter used for each ROC curve is produced by simulating a zero mean complex Gaussian process whose covariance matrix is specified, and so is not regressed from real data. This is then multiplied by a simulation of S with density (28) to produce a simulation of the SIRP.

4.3.1 Examples for horizontal polarisation

Two cases are considered for the horizontally polarised case. The first is for the scenario where the KK-Distributed clutter takes parameters $c_1 = 1, c_2 = 3.27, \bullet = 4.158$, and $k = 0.01$. These parameters have been selected based upon the estimates for the clutter set run34683 described above, and with parameters estimated based upon the results in Dong (2006). For this case, the azimuth angle is $225°$, which is nearly upwind. The number of looks is $N = 30$, and the clutter covariance matrix has been produced using simulated variables. Additionally, for this example, the normalised Doppler frequency is generated from $f_D = 1$ and $T_s = 0.5$. The corresponding ROC curve can be found in Figure 2 (left subplot). This shows the optimal detector and GLRT detector matching very closely. Increasing the number of Monte Carlo samples improves the plot, but takes long periods to generate. The main feature one can observe from this plot is that the GLRT performs very well in this case, and certainly outperfoms the WMF.

The second example considered is illustrated also in Figure 2 (right subplot). This is for the case where the KK-Distribution has parameters $c_1 = 8, c_2 = 46.16, \bullet = 4.684, k = 0.01, f_D = 0.8, T_s = 0.5$, with the number of looks $N = 20$. The clutter parameters have been estimated from data set run34683 with azimuth angle $190°$. The clutter covariance matrix was generated with random Gaussian numbers. The plot shows the same phenomenon as for the previous example.

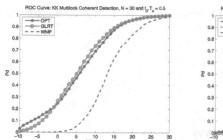
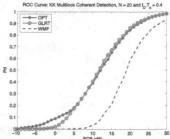

Fig. 2. ROC curves, Horizontal polarisation, clutter set run34683, azimuth angles 225° (left plot) and 190° (right plot). OPT corresponds to (33), GLRT to the suboptimal rule using (35) and WMF is (39).

4.3.2 Examples for vertical polarisation

The vertically polarised case is illustrated in Figure 3. The first subplot is for the scenario where $c_1 = 25, c_2 = 26.5, \bullet = 8.315, k = 0.01, f_d = 100$ and $T_s = 0.5$. $N = 10$ looks have been used. These correspond to run34690, with azimuth angle 45°, which is almost down wind in direction. For this case, it is interesting to note that the GLRT and WMF match very closely. It is only under increased magnification that it can be shown the GLRT is marginally better. The same phenomenon is also demonstrated for the case where $c_1 = c_2 = 35, \bullet = 12.394, f_D = 10$ and $T_s = 0.5$, with $N = 5$ looks. This is illustrated in Figure 3 (right subplot). This example is based upon the clutter set run34690, at an azimuth angle of 315°, which is approximately crosswind.

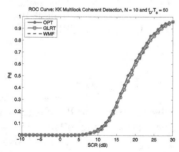
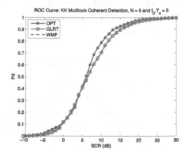

Fig. 3. ROC curves corresponding to the vertically polarised case, for clutter set run34690, azimuth angles 45° (left plot) and 315° (right plot). Here, the GLRT and WMF match almost exactly, suggesting the WMF is a suitable suboptimal decision rule.

4.3.3 Analysis of detectors

Examination of ROC curves for other Ingara clutter sets showed roughly the same results. For the vertically polarised channel, the WMF was a valid approximation to the GLRT, and is preferable because it does not require knowledge of clutter parameters. This result can be explained from the fact that the vertically polarised clutter is not as spiky as the horizontally polarised case, and the Gaussian distribution is the limit of a K-Distribution as the shape parameter increases. The latter results in less spiky clutter. For the horizontally polarised case, the clutter is spikier and so the GLRT is much better.

5. Conclusions and further research

The KK-Distribution was embedded within a SIRP in the complex domain, so that the Neyman-Pearson optimal detector could be constructed for multilook detection. A GLRT solution was then constructed, and this, together with the WMF suboptimal decision rule, were compared using ROC curves. These ROC curves were generated using clutter parameter estimates based upon the Ingara data set. For the horizontally polarised case, it was shown that the GLRT produced a very good probability of detection for a Swerling 1 target model. In the case of vertically polarised returns, the WMF often matched the performance of the GLRT. Hence, in such cases, the WMF can be used as a simpler suboptimal detector.

Further work will be spent on trying to find suboptimal decision rules based upon simplification of the GLRT solution. Additionally, performance analysis of detectors for other target models will be undertaken.

6. References

Barnard, T. J. and Khan, F. (2004). Statistical Normalization of Spherically Invariant Non-Gaussian Clutter. IEEE J. Ocean. Eng. Vol. 29, 303-309.

Beaumont, G. P. (1980). Intermediate Mathematical Statistics, Chapman and Hall, London.

Conte, E. and Longo, M. (1987). Characterisation of radar clutter as a spherically invariant random process. Proceed. IEEEVol. 134 F, 191-197.

Conte, E., Lops, M. and Ricci, G. (1995). Asymptotically Optimum Radar Detection in Compound Gaussian Clutter. IEEE Trans. Aero. Elec. Sys. Vol. 31, 617-625.

Crisp, D. J., Rosenberg, L., Stacy, N. J. and Dong, Y. (2009). Modelling X-Band Sea Clutter with the K-Distribution: Shape Parameter Variation, IEEE Conf. Surveillance for a Safer World, Bordeaux, France.

Crombie, D. (1955). Doppler Spectrum of Sea Echo at 13.56Mc/s. Nature Vol. 175, 681-682.

Dong, Y. (2006). Distribution of X-Band High Resolution and High Grazing Angle Sea Clutter. DSTO Research Report DSTO-RR-0316.

Durrett, R. (1996) Probability: Theory and Examples, 2nd Edition. Wadsworth, California.

Evans, M., Hastings, N. and Peacock, B.(2000). Statistical Distributions, 3rd Edition. Wiley, New York.

Farina, A., Gini, F., Greko, M. V. and Lombardo, P. (1995). Coherent Radar Detection of Targets Against a Combination of K-Distributed and Gaussian Clutter. Proceed. IEEE Int. Radar, Conf., 83-88.

Gini, F., Greko, M. V., Farina, A. and Lombardo, P. (1998) Optimum and Mismatched Detection Against K-Distributed Plus Gaussian Clutter. IEEE Trans. Aero. Elec. Sys. Vol. 34, 860-876.

Jakeman, E. and Pusey, P. N. (1976). A model for Non-Rayleigh Sea Echo. IEEE Trans. Antennas Prop. Vol AP24, 806-814.

Jakeman, E. and Pusey, P. N. (1977). Statistics of Non-Rayleigh Sea Echo. IEE Conf. Publ. 155 Radar 1977, 105-109.

Kreyszig, E. (1978). Introductory Functional Analysis with Applications. Wiley, New York.

Levanon, N. (1988). Radar Principles. Wiley, New York.

Mahafza, B. R.(1998). Introduction to Radar Analysis. CRC Press, Florida.

Middleton, D. (1999). New Physical-Statistical Methods and Models for Clutter and Reverberation: The KA-Distribution and Related Probability Structures. IEEE J. Oceanic Eng.Vol 24, 261-284.

Neyman, J. and Pearson, E. (1933). On the Problem of the Most Efficient Tests of Statistical Hypotheses. Phil. Trans. Royal Soc. Lond. Series A 231, 289-337.

Peebles, P. Z. (1998). Radar Principles. Wiley, New York.

Rangaswamy, M., Weiner, D. and Ozturk, A. (1991). Simulation of Correlated Non-Gaussian Interference for Radar Signal Detection, Proceed. IEEE, 148-152.

Rangaswamy, M, Weiner, D, and Ozturk, A. (1993), Non-Gaussian Random Vector Identification Using Spherically Invariant Random Processes. IEEE Trans. Aero. Elec. Sys. Vol. 29, 111-123.

Rosenberg, L., Crisp, D. J. and Stacy, N. J. (2010). Analysis of the KK-Distribution with Medium Grazing Angle Sea Clutter. IET Radar, Sonar, Navig. Vol. 4, 209-222.

Schleher, D. C. (1976). Radar Detection in Weibull Clutter. IEEE Trans. Aero. Elec. Sys. Vol. AES-12, 736-743.

Shnidman, D. A. (1999). Generalized Radar Clutter Model. IEEE Trans. Aero. Elec. Sys. Vol.35, 857-865.

Skolnik, M. (2008). Introduction to Radar Systems: Third Edition. McGraw-Hill, New York.

Stacy, N., Badger, D., Goh, A., Preiss, M. and Williams, M. (2003). The DSTO Ingara Airborne X-Band SAR Polarimetric Upgrade: First Results. IEEE Proceed. Int. Symp. Geo. Remote Sensing (IGARSS) Vol. 7, 4474-4475.

Stacy, N, Crisp, D., Goh, A., Badger, D. and Preiss, M. (2005). Polarimetric Analysis of Fine Resolution X-Band Sea Clutter Data. Proc. IGARSS 05, 2787-2790.

Stimson, G. W. (1988). Introduction to Airborne Radar: Second Edition. Scitech Publishing, Inc, Rayleigh, NC.

Trunk, G. V. and George, S. F. (1970). Detection of Targets in Non-Gaussian Sea Clutter. IEEE Trans. Aero. Elec. Sys. Vol.AES-6, 620-628.

Valeyrie, N., Garello, R., Quellec, J.-M. and Chabah, M. (2009). Study of the Modeling of Radar Sea Clutter Using the KA-Distribution and Methods for Estimating its Parameters, Radar Conf: Surveillance for a Safer World.

Ward, K. D. (1981). Compound Representation of High Resolution Sea Clutter. IEE Elec. Lett. Vol. 17, 561-563.

Ward, K. D., Watts. S. and Tough, R. J. A. (2006). Sea Clutter: Scattering, the K-Distribution and Radar Performance. IET Radar, Sonar and Nav., Series 20, IET London.

Ward, K. D. and Tough, R. J. A. (2002). Radar Detection Performance in Sea Clutter with Discrete Spikes. International Radar Conference, 253-257.

Watts, S., Ward, K. D. and Tough, R. J. A. (2005). The Physics and Modelling of Discrete Spikes in Radar Sea Clutter. Proceedings of International Radar Conference.

Wise, G. and Gallagher, N. C.(1978). On Spherically Invariant Random Processes. IEEE Trans. Info. Theory Vol. IT-24, 118-120.

Yao, K. (1973). A Representation Theorem and Its Application to Spherically-Invariant Random Processes. IEEE Trans. Info. Theory Vol. IT-19, 600-608.

8

Application of the Mode Intermittent Radiation in Fading Channels

Mihail Andrianov[1,2] and Igor Kiselev[3]

*[1]Institute of Radio-Engineering and Electronics of Russian
Academy of Science (Kotelnirov's Institute)
[2]Ltd Research & Development Company "Digital Solution"
[3]OAO "Network Technologies"
Russia*

1. Introduction

The effectiveness of radio lines which use radio signals from deep fading signal levels can be increased by the use of transmitting devices to adapt to changing the radio channel's parameters. To adapt to the conditions of radio propagation and effects of interfering signals are used the system with: automatic adjustment of transmitter power; automatic variable speed transmission; adaptation in frequency, switching modulation types, etc. To improve noise immunity can be used as the mode of intermittent radiation or intermittent radio communications. Intermittent signal transmission principle provides the highest level of agreement when using this mode in one direction of the double-side radio line.

For a long time, almost until the end of the XX-th century, transmission of radio signals on double-side radio line was carried out on the same principles and with the same energy potential in the forward and reverse directions. Such lines are called symmetric. Currently being developed and are used asymmetric radio lines, in which the transmission of signals in forward and reverse directions is organized on different principles and with different energy potentials. There are a variety of options for the construction of forward and reverse directions.

In constructing the double-sided asymmetric radio lines in either direction can be used by the mode of intermittent radiation. The principal feature of this option is the ability to completely turn off the transmitter during the fading signals and thus reduce the average output power.

In the well-known radio lines of intermittent communication (for example meteor communication), when an intermittent communication is carried out in both directions is not possible to turn off the transmitter. Interruption of communication in both directions leads to the necessity at all times to transmit sounding signals and completely turn the transmitter on the break of communication is not possible. There are other technical difficulties in the organization of bilateral intermittent communication.

More simply a double-side line will created when using intermittent communications in one direction of communication, especially when using technology TDD (Time Divide Duplex).

In the case of TDD fading signal levels at the receiver inputs at opposite ends of the radio lines are correlated. This allows measuring the signal level at the input of the receiver to estimate the signal level at the other end, and thus dispense with a special back channel communications. In recent decades, great importance is the communication with mobile objects. For such objects, it is often important to minimize the average power consumption of the power supply and increase the duration of their work. From this perspective, the opportunity to switch off the transmitter while fading, and not just to stop the transmission of information signals is an important factor.

Intermittent mode radiation can be used in combination with other ways to improve radio communications. The development of radio engineering has led to what is now practically all the possible ways of improving known and well understood, therefore, appears to be the most significant results to further improve the radio communication systems can be obtained by complexation of different ways. Sometimes there are situations when the use of one method for increasing the efficiency entails reducing the effectiveness of another, but by introducing restrictions on the parameters of signals and separate units of equipment able to maintain the effectiveness of both methods.

It is interesting to consider the use of complex methods diversity of signals and intermittent of radiation. The use of diversity transfer mode reduces the efficiency of intermittent radiation. However, for small number branches of the diversity efficiency of intermittent radiation mode remains significant. There are several options complexation mode of intermittent radiation and diversity signal. In essence, the intermittent mode radiation can be combined with different variants of MIMO (Multiply Input Multiply Output), the use of which is one approach to improving the system of broadcast signals. A number of advances in microelectronics, the possibility of solutions to complex analytical problems have led to many versions of MIMO schemes are implemented in telecommunication systems.

Earlier studies performed immunity intermittent communication concerned only cases receiving single and Rayleigh fading. Similar studies with other statistics of fading weren't made. Study the effectiveness of intermittent connection when its application with the diversity of signals is unknown to authors.

In this chapter the noise immunity of transmission digital communications in fading channels is analyzed in case using the intermittent radiation.

2. The analysis of noise immunity radio-line at usage mode intermittent of radiation

2.1 Error probably of signal reception, use factor of radio-line when usage intermittent radio communication channel with different types of fading

The main parameter that characterizes the noise immunity of digital communication systems is the probability of error. Because in considered channels of communication signal propagation suffers fading, the probability of error is a random variable, and the quality of communication is estimated by the mean probability of error, which is usually defined as mathematical expectation this random variable. Using the known expression, determining the probability of error for noncoherent reception of orthogonal signals in channels with Gaussian noise and the probability density function of the signal / noise ratio (SNR), was found analytic expressions for the dependence of the average probability of error from the mean SNR and the level of threshold's interruption for the most commonly used when describing the fading distribution laws of random variables:

$$P_{nc}(\gamma_0) = \frac{1}{2 \cdot \eta(\gamma_0)} \int_{\gamma_t}^{\infty} \exp(-\alpha\gamma) \cdot \frac{1}{\gamma_0} \cdot \exp\left(-\frac{\gamma}{\gamma_0}\right) d\gamma = \frac{1}{2 \cdot \eta(\gamma_0)} \frac{\exp\left[-\alpha\gamma_t\left(1+\frac{1}{\alpha\gamma_0}\right)\right]}{1+\alpha\gamma_0}, \tag{1}$$

$$P_{nc}(\gamma_0) = \frac{1}{2\eta(\gamma_0)} \int_{\gamma_t}^{\infty} \exp(-\alpha\gamma) \frac{1}{\gamma_0 - \gamma_a} \cdot \exp\left(-\frac{\gamma + \gamma_a}{\gamma_0 - \gamma_a}\right) \cdot I_0\left(\frac{2 \cdot \sqrt{\gamma \cdot \gamma_a}}{\gamma_0 - \gamma_a}\right) d\gamma, \tag{2}$$

$$P_{nc}(\gamma_0) = \frac{1}{4\eta(\gamma_0)\sqrt{\pi}} \int_{\gamma_t}^{\infty} \frac{1}{\gamma} \cdot \frac{1}{\sqrt{\ln\gamma_0 - \ln\gamma_{0\_ref}}} \exp\left[-\frac{\left(\ln\sqrt{\frac{\gamma}{\gamma_{0\_ref}}}\right)^2}{\ln\gamma_0 - \ln\gamma_{0\_ref}}\right] \cdot e^{-\alpha\gamma} d\gamma, \tag{3}$$

$$P_{nc}(\gamma_0) = \frac{1}{2\eta(\gamma_0)} \frac{1}{\Gamma(m)} \left[\frac{m}{\gamma_0 \cdot \eta(\gamma_0)}\right]^m \int_{\gamma_t\eta(\gamma_0)}^{\infty} \gamma^{m-1} \exp\left[-\left(\frac{m}{\gamma_0 \cdot \eta(\gamma_0)} + \alpha\right)\gamma\right] d\gamma =$$

$$= \frac{1}{2} \frac{1}{\Gamma\left(m,\frac{\gamma_t}{\gamma_0}m\right)} \left[1 + \alpha\frac{\gamma_0}{m} \frac{\Gamma\left(m,\frac{\gamma_t}{\gamma_0}m\right)}{\Gamma(m)}\right]^{-m} \left[\Gamma\left(m, m\frac{\gamma_t}{\gamma_0} + \alpha\gamma_t \frac{\Gamma\left(m,\frac{\gamma_t}{\gamma_0}m\right)}{\Gamma(m)}\right)\right], \tag{4}$$

where analytical expressions (1-4) define mean values probability of error under the laws accordingly Rayleigh, the generalized Rayleigh, lognormal and Nakagami, m-parameter of fading and $m> 0.5$, α - the constant coefficient equal 0.5 and 1 for accordingly frequency and phase demodulations, γ, γ_0, γ_t – accordingly current, average and threshold value SNR, $\eta(\gamma_0)$ - the parameter named use factor (utilization factors or fill factor) of a radio-line, $\Gamma(m) = \int_0^{\infty} t^{m-1}e^{-t}dt$ and $\Gamma(m,n) = \int_n^{\infty} t^{m-1}e^{-t}dt$ accordingly gamma-function and incomplete gamma-function. Here and below value α will make 1.

Integral in the formula (4) is tabular, defined by expression

$$\int_u^{\infty} x^{\nu-1} \exp(-\mu x) dx = \frac{\Gamma(\nu, \mu \cdot u)}{\mu^\nu}. \tag{5}$$

Value γ_a correspond SNR for the regular component of a signal when a signal fading down under laws of the generalized Rayleigh (formula 2). When a signal fading under the lognormal law (formula 3) value γ_{0_ref} correspond SNR in the nonperturbed environment.

Random changes of an envelope of a signal of millimeter wave range at a lognormal fading, unlike a fading under Rayleigh, the generalized Rayleigh and Nakagami laws, are caused not by its total interference components, and fluctuation of dielectric transmittivity ε, a wave refraction index n owing to turbulence of troposphere.

Expression $\eta(\gamma_0)$ defines use factor radio-line at intermittent communication (the relation of transmission time of the data to the general time of a communication session) and its lowering corresponds to reduction of spectral efficiency, accordingly for the above-stated types of a fading will make, according to analytical expressions (6-9)

$$\eta(\gamma_0) = \int_{\gamma_t}^{\infty} \frac{1}{\gamma_0} \cdot \exp\left(-\frac{\gamma}{\gamma_0}\right) d\gamma = \exp\left(-\frac{\gamma_t}{\gamma_0}\right), \tag{6}$$

$$\eta(\gamma_0) = \int_{\gamma_t}^{\infty} \frac{1}{\gamma_0 - \gamma_a} \cdot \exp\left(-\frac{\gamma + \gamma_a}{\gamma_0 - \gamma_a}\right) \cdot I_0\left(\frac{2 \cdot \sqrt{\gamma \cdot \gamma_a}}{\gamma_0 - \gamma_a}\right) d\gamma, \tag{7}$$

$$\eta(\gamma_0) = \int_{\gamma_t}^{\infty} \frac{1}{2\sqrt{\pi}} \cdot \frac{1}{\gamma} \cdot \frac{1}{\sqrt{\ln\gamma_0 - \ln\gamma_{0\_ref}}} \exp\left[-\frac{\left(\ln\sqrt{\frac{\gamma}{\gamma_{0\_ref}}}\right)^2}{\ln\gamma_0 - \ln\gamma_{0\_ref}}\right] d\gamma = \frac{1}{2} \cdot erfc\left(\frac{1}{2} \cdot \frac{\ln\gamma_t - \ln\gamma_{0\_ref}}{\sqrt{\ln\gamma_0 - \ln\gamma_{0\_ref}}}\right), \tag{8}$$

$$\eta(\gamma_0) = \frac{1}{\Gamma(m)}\left(\frac{m}{\gamma_0}\right)^m \int_{\gamma_t}^{\infty} \gamma^{m-1} \exp\left(\frac{\gamma \cdot m}{\gamma_0}\right) = \frac{\Gamma\left(m, m\frac{\gamma_t}{\gamma_0}\right)}{\Gamma(m)}, \tag{9}$$

where $erfc(x) = \frac{2}{\sqrt{\pi}} \int_{x}^{\infty} \exp\left(-t^2\right) dt$ - error function complement. Integral in the formula (9) is tabular (5). Analytical expression (6) is a special case of the formula (9) at $m=1$.

Let's accept $\gamma_t = k\gamma_0$, $(0<k<\infty)$ and value k is fixed, then the use factor of a radio- line in case of the Rayleigh fading is fixed, doesn't depend from γ_0. In cases of a fading of a signal under laws of the generalized Rayleigh, lognormal and Nakagami it depends accordingly on ratios k, γ_a, γ_0; k, γ_{0_ref}, γ_0 and k, m, and in two last cases the value of the specified coefficient will be defined by formulas

$$\eta(\gamma_0) = \frac{1}{2} \cdot erfc\left[\frac{1}{2} \cdot \left(\frac{\ln k}{\sqrt{\ln\frac{\gamma_0}{\gamma_{0\_ref}}}} + \sqrt{\ln\frac{\gamma_0}{\gamma_{0\_ref}}}\right)\right]; \quad \eta(m) = \frac{\Gamma(m, km)}{\Gamma(m)}.$$

Analyzing dependences of probability of errors on mean value SNR, it is necessary to consider that at constant instantaneous power of the transmitter at intermittent communication energy of bit of a signal decreases proportionally to use factor of a radio-line. Therefore for, the comparative analysis of noise immunity at intermittent communication and without it, it is expedient in the first case at calculation of probability of error accordingly to reduce level of a threshold and an average of value SNR and that is showed in analytical expression (4). In the absence of it, for correct comparing of the specified probabilities of errors, it is necessary to admit that instantaneous power transmitters at intermittent communication comparing communication without interruption is increased in inverse proportion to use factor of radio-line that is illustrated in formulas (1-3).

Application algorithm of intermittent communication reduces probabilities of errors in the channel with a fading and, especially with growth of mean value SNR (γ_0) at some lowering of spectral efficiency (use factor of a radio-line). For example, for the Rayleigh channel (1) at $\gamma_t = \gamma_0$; $\eta(\gamma_0)=1/e$, (6) therefore

$$P_{nc}(\gamma_0) = \frac{e}{2} \int_{\gamma_0}^{\infty} \exp(-\alpha\gamma) \cdot \frac{1}{\gamma_0} \cdot \exp\left(-\frac{\gamma}{\gamma_0}\right) d\gamma = \frac{1}{2} \frac{\exp(-\alpha\gamma_0)}{1+\alpha\gamma_0} = \frac{1}{2}\frac{1}{1+\alpha\gamma_0} \cdot \exp(-\alpha\gamma_0).$$

Thus, application of intermittent communication with a threshold equal γ_0 reduces probability of error incoherent reception in comparing probability of errors in the Rayleigh channel without interruption exponentially, in $\exp\alpha\gamma_0$ time.

2.2 Comparing of energetic effectiveness usage of intermittent communication and noise immunity coding

The comparative analysis of energetic efficiency of the radio-line using intermittent communication in comparison with a radio-line, using the noise immunity code for the channel with a fading of Nakagami is carried out. Variation of coefficient k we will select for $m = 1.4$ such value of a threshold ($\gamma_t = k\gamma_0$) that for them η (9) it was equal 0.5.

Comparing of effectiveness of communication is expedient for carrying out on energetic coding gain and energetic gain of intermittent communication at the given probabilities of errors. Energetic coding gain for intermittent communication ($m =1.4$; $\eta=0.5$) and the convolutional code of speed 0.5 are presented to tab. 1. At realization of intermittent communication, as well as transmission of an encoded signal, energy of the bit (average power) at the fixed instantaneous power of the transmitter will decrease twice, the width of spectra of both signals thus will double, and spectral efficiency accordingly twice will decrease.

The given probabilities of error	10^{-3}	10^{-4}	10^{-5}	10^{-6}
Energetic gain of coding (dB)	4	5,1	5,7	6
Energetic gain of intermittent communication (dB)	11,3	16,1	22,7	28,9

Table 1.

The data presented to tab. 1 show that intermittent communication, is more effective comparing convolutional code usage (at equal values of use factor of radio-line and speed of coding). Besides, in the channel with a fading for possibility of realization of decoding of the convolutional code, originating burst errors necessary to convert in single-error. Normally it is fulfilled by operations of interleaving/de-interleaving which are implemented with additional hardware expenses that causes additional signal delays. At deeper fading of a signal when $0.5 <m <1$ energetic gain in application of intermittent communication will increase.

3. Complexation of intermittent communication and diversity reception

3.1 Substantiation of expediency of complexation of intermittent communication and diversity reception of signals

It is known that at use of diversity reception fading in the channel becomes less deep, due it increases the noise immunity of signal reception. At complexation intermittent communication with diversity reception, gain comparing to diversity reception without intermittent communication will be less than that of an intermittent connection, comparatively the continuous reception in a single. And this gain decreases with magnification of number of branches of diversity as with growth of the last depth of a fading decreases. However usage of intermittent communication at restricted number of branches of the diversity, normally put into practice, leads to noticeable magnification of noise immunity. Besides, application of diversity reception at intermittent communication increases use factor of radio-line that also should raise noise immunity of reception and spectral efficiency of data transfer.

Value η_M, named use factor of radio-line, essentially influences noise immunity of reception of signals. At complexation of intermittent communication and diversity reception in communication paths for calculation probability of error reception it is necessary to define use factor of radio-line at various number of branches of diversity (M) depending on γ_0.

3.2 Analysis of noise immunity at complexation of intermittent communication and diversity reception with a combination branches of diversity on algorithm of an auto select

It is known that the probability density function at M - the multiple diversity reception with an ideal auto select at an independent homogeneous fading can be defined from expression

$$W_M(\gamma) = \frac{d\left(\int_0^\gamma f_\gamma \, d\gamma\right)^M}{d\gamma},$$
(10)

where M is number of branches of diversity, f_γ - probability density SNR at various types of a fading.

At complexation of intermittent communication and diversity reception with a combination of branches of diversity on algorithm of an auto select, use factor of radio-line at various number of branches of diversity (M) depending on γ_0 are defined according to expression

$$\eta_M(\gamma_0) = \int_{\gamma_t}^{\infty} W_M(\gamma)\,d\gamma = \int_{\gamma_t}^{\infty} \frac{d\left(\int_0^{\gamma} f_\gamma\,d\gamma\right)^M}{d\gamma}\,d\gamma \ . \tag{11}$$

Use factor of radio-line at complexation of intermittent communication and diversity reception with a combination of branches of diversity on algorithm of an auto select at a signal fading under accordingly Rayleigh (12), the generalized Rayleigh (13), lognormal (14) and Nakagami (15) laws will be defined according to (11) expressions

$$\eta_M(\gamma_0) = \int_{\gamma_t\cdot}^{\infty} \frac{d\left[1-\exp\left(-\dfrac{\gamma}{\gamma_0}\right)\right]^M}{d\gamma}\,d\gamma = 1-\left[1-\exp\left(-\dfrac{\gamma_t}{\gamma_0}\right)\right]^M, \tag{12}$$

$$\eta_M(\gamma_0) = \int_{\gamma_t}^{\infty} \frac{d\left[\int_0^{\gamma} \dfrac{1}{\gamma_0-\gamma_a}\cdot\exp\left(-\dfrac{\gamma+\gamma_a}{\gamma_0-\gamma_a}\right)\cdot I_0\left(\dfrac{2\sqrt{\gamma_a\cdot\gamma}}{\gamma_0-\gamma_a}\right)d\gamma\right]^M}{d\gamma}\,d\gamma, \tag{13}$$

$$\eta_M(\gamma_0) = \int_{\gamma_t}^{\infty} \frac{d\left[\dfrac{1}{2}+\dfrac{1}{2}erf\left(\dfrac{1}{2}\cdot\dfrac{\ln\gamma-\ln\gamma_{0\_ref}}{\sqrt{\ln\gamma_0-\ln\gamma_{0\_ref}}}\right)\right]^M}{d\gamma} =$$

$$= 1-\left[\dfrac{1}{2}+\dfrac{1}{2}\cdot erf\left(\dfrac{1}{2}\cdot\dfrac{\ln\gamma_t-\ln\gamma_{0\_ref}}{\sqrt{\ln\gamma_0-\ln\gamma_{0\_ref}}}\right)\right]^M \tag{14}$$

$$\eta_M(\gamma_0) = \frac{M}{\Gamma(m)}\left(\frac{m}{\gamma_0}\right)^m \int_{\gamma_t}^{\infty}\left[1-\frac{\Gamma\left(m,\dfrac{\gamma}{\gamma_0}m\right)}{\Gamma(m)}\right]^{M-1}\exp\left(-\frac{\gamma}{\gamma_0}m\right)\gamma^{m-1}d\gamma =$$

$$= 1-\left[\frac{\Gamma(m)-\Gamma\left(m,\dfrac{\gamma_t}{\gamma_0}m\right)}{\Gamma(m)}\right]^M \tag{15}$$

where $erf(x) = \dfrac{2}{\sqrt{\pi}}\int_0^x \exp\left(-t^2\right)dt$ error function.

Analytical expression (12) is a special case of the formula (15) at m, equal 1. Dependence of use factor of radio-line on number of branches of diversity for a fading under the law of Nakagami, at $\gamma_t=\gamma_0$ and m equal 0.7, under the formula (15), is presented on fig. 1.

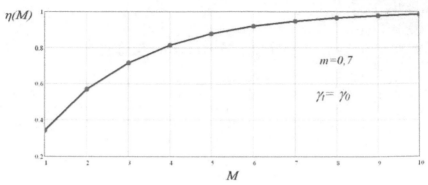

Fig. 1. Dependence of use factor of radio-line from M for a fading under the law of Nakagami, at $\gamma_t = \gamma_0$ and m equal 0.7.

Similarly, methods of calculation for a single reception and intermittent communication is defined depending on the average error probability of noncoherent reception in the complexation intermittent links and diversity reception of the average SNR (γ_0) for the cases of the fading signal differential binary phase shift key (D-BPSK), respectively, under the laws of the Rayleigh (16) [], the generalized Rayleigh (17) [], log-normal (18) [] and Nakagami (19)

$$P_M(\gamma_0) = \frac{1}{2\eta_M} \cdot \int_{\gamma_t \cdot \eta_M(\gamma_0)}^{\infty} \frac{d\left[1 - \exp\left(-\dfrac{\gamma}{\gamma_0}\right)\right]^M}{d\gamma} \cdot \exp(-\alpha\gamma)d\gamma =$$

$$\frac{1}{2\eta_M(\gamma_0)} \cdot M \cdot \exp\left[-\alpha\gamma_t \cdot \eta_M(\gamma_0)\right] \cdot \left[\sum_{k=1}^{M} C_{M-1}^{k-1} \cdot \frac{(-1)^{k-1} \cdot \exp\left(-\dfrac{k\gamma_t}{\gamma_0}\right)}{\alpha\gamma_0 \cdot \eta_M(\gamma_0) + k}\right]'$$

(16)

$$P_M(\gamma_0) = \frac{1}{2\eta_M(\gamma_0)} \int_{\gamma_t}^{\infty} \exp(-\alpha\gamma) \frac{d\left[\int_0^{\gamma} \dfrac{1}{\gamma_0 - \gamma_a} \cdot \exp\left(-\dfrac{\gamma + \gamma_a}{\gamma_0 - \gamma_a}\right) \cdot I_0\left(\dfrac{2\sqrt{\gamma_a \cdot \gamma}}{\gamma_0 - \gamma_a}\right)d\gamma\right]^M}{d\gamma} d\gamma ,$$

(17)

$$P_M(\gamma_0) = \frac{1}{\eta_M} \int_{\gamma_t \eta_M}^{\infty} M \cdot \left\{ \frac{\left[\dfrac{1}{2} + \dfrac{1}{2} \cdot erf\left(\dfrac{1}{2}\dfrac{\ln\gamma - \ln\gamma_{0\_ref}\eta_M}{\sqrt{\ln\gamma_0 - \ln\gamma_{0\_ref}}}\right)\right]^{M-1}}{2\sqrt{\pi}} \cdot \frac{\exp\left[-\dfrac{1}{4}\dfrac{\left(\ln\gamma - \ln\gamma_{0\_ref}\eta_M\right)^2}{\ln\gamma_0 - \ln\gamma_{0\_ref}}\right]}{\gamma \cdot \sqrt{\ln\gamma_0 - \ln\gamma_{0\_ref}}} \right\} \frac{\exp(-\alpha\gamma)}{2} d\gamma ,$$

(18)

$$P_M(\gamma_0) = \frac{M}{2\eta_M(\gamma_0)} \left[\frac{m}{\gamma_0\eta_M(\gamma_0)}\right]^m \frac{1}{\Gamma(m)} \int_{\gamma_t\eta_M(\gamma_0)}^{\infty} \left\{1 - \frac{\Gamma\left[m, \dfrac{m\gamma}{\gamma_0\eta_M(\gamma_0)}\right]}{\Gamma(m)}\right\}^{M-1} \gamma^{m-1} \exp\left\{-\left[\frac{m}{\gamma_0\eta_M(\gamma_0)} + \alpha\right]\gamma\right\}d\gamma .$$

(19)

Dependences of probability of errors at a signal fading under the law of Nakagami, and complexation of intermittent communication and diversity reception with a combination of branches of diversity on algorithm of an auto select, at $\gamma_t = \gamma_0$ and at m equal 0.7 in comparison with continuous diversity reception are shown in Fig. 2.

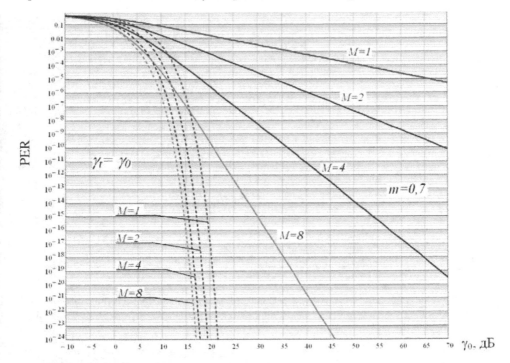

Fig. 2. Probabilities of errors at complexation of intermittent communication and diversity reception, with a combination of branches of diversity on algorithm of an auto select for a signal fading under the law of Nakagami at $\gamma_t = \gamma_0$ and at m equal 0.7 in comparison with continuous diversity by reception (continuous curves) at various number of branches of diversity (M).

Dependence of probability of an error at a lognormal fading of a signal (18) is obtained for a non stationary case, when the dispersion of the lognormal process defined by a ratio at fixed $\gamma_{0\_ref.}$ isn't a constant

$$2\sigma_\chi^2 = \ln\langle X^2 \rangle = \ln \gamma_0 - \ln \gamma_{0\_ref.} = \ln \frac{\gamma_0}{\gamma_{0\_ref.}}. \tag{20}$$

In practice, at calculation of noise immunity of communication lines the dispersion of lognormal process normally is the fixed value. For this case it is expedient to define analytical expressions of probability density SNR (21), use factor of radio-line (22, 23) and probabilities of error reception of incoherent signal differential D-BPSK (24) in the conditions of a lognormal fading at complexation of intermittent communication and diversity reception with a combination branches of diversity on algorithm of an auto select. The diagrams illustrating specified dependences are presented accordingly on fig. 3, 4, 5. Analytical expression of probability density SNR for the specified type of a fading is defined by a method of symbolic mathematical modeling in the environment of MathCAD 14

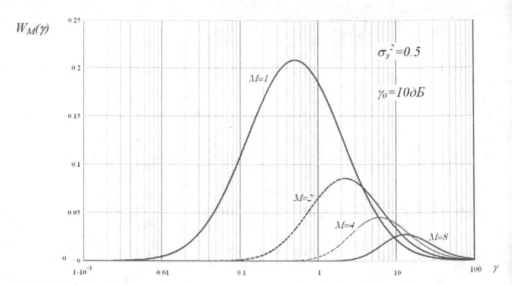

Fig. 3. Probability densities SNR diversity reception of signals in the channel with a lognormal fading at combination of branches of diversity on algorithm of an auto select, at the fixed dispersion (σ_χ^2) and mean value SNR (γ_0).

$$W_M(\gamma) = \frac{M}{2} \cdot \frac{1}{\sqrt{2\pi\sigma_\chi^2}\gamma} \cdot \left[\frac{1}{2} + \frac{1}{2} \cdot erf\left(\frac{\sqrt{2}}{2} \cdot \frac{\ln\sqrt{\frac{\gamma}{\gamma_0}} + \sigma_\chi^2}{\sigma_\chi} \right) \right]^{M-1} \cdot \exp\left[-\frac{\left(\ln\sqrt{\frac{\gamma}{\gamma_0}} + \sigma_\chi^2 \right)^2}{2\sigma_\chi^2} \right] =$$

$$= \frac{M}{2} \cdot \frac{1}{\sqrt{2\pi\sigma_\chi^2}\gamma} \cdot \left[\Phi\left(\frac{\ln\sqrt{\frac{\gamma}{\gamma_0}} + \sigma_\chi^2}{\sigma_\chi} \right) \right]^{M-1} \cdot \exp\left[-\frac{\left(\ln\sqrt{\frac{\gamma}{\gamma_0}} + \sigma_\chi^2 \right)^2}{2\sigma_\chi^2} \right], \qquad (21)$$

the use factor of radio-line will thus be defined under the formula

$$\eta_M(\gamma_0) = \int_{\gamma_t}^{\infty} W_M(\gamma)d\gamma = 1 - \left[\frac{1}{2} + \frac{1}{2}erf\left(\frac{1}{\sqrt{2}} \frac{\ln\sqrt{\frac{\gamma_t}{\gamma_0}} + \sigma_\chi^2}{\sigma_\chi} \right) \right]^M = 1 - \left[\Phi\left(\frac{\ln\sqrt{\frac{\gamma_t}{\gamma_0}} + \sigma_\chi^2}{\sigma_\chi} \right) \right]^M . \qquad (22)$$

At $\gamma_t = \gamma_0$ the use factor of radio-line depends only on number branches of diversity, fig. 4

$$\eta_M = 1 - \left[\frac{1}{2} + \frac{1}{2}erf\left(\frac{\sigma_\chi}{\sqrt{2}}\right)\right]^M = 1 - \left[\Phi(\sigma_\chi)\right]^M,\qquad(23)$$

where $\Phi(x)$ is normal cumulative distribution function.

The probability of error reception for signal D-BPSK will be defined thus according to expression

$$P_M(\gamma_0) = \frac{1}{\eta_M}\int\limits_{\gamma_0\eta_M}^{\infty} W_M(\gamma)\cdot\frac{\exp(-\alpha\gamma)}{2}d\gamma.\qquad(24)$$

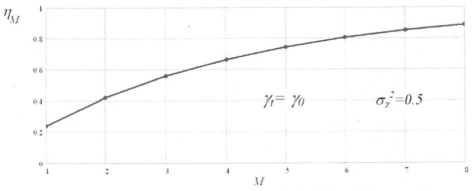

Fig. 4. Dependence of use factor of radio-line from number branches of diversity (M) at $\gamma_t = \gamma_0$ and the fixed dispersion (σ_χ^2).

In the fading signal under the laws of Nakagami and lognormal use factor of the radio line increases, respectively, with increasing parameter of fading (m) and lowering variance ((σ_χ^2)). The probabilities of errors are reduced under these conditions.

3.3 The noise immunity analysis at complexation of intermittent communication with diversity reception and optimal addition of branches diversity

Let's consider a variant of diversity reception of signal D-BPSK when combination of branches of diversity is carried out by a method of optimal addition for which the equality defined by expression is valid

$$P_M(\gamma_\Sigma) = \frac{1}{2}\exp\left(-\sum\limits_{i=1}^{M}\gamma_i\right),\qquad(25)$$

where value SNR of optimally added signal (γ_Σ) will be to equally arithmetical total SNR of all of i channels.

It is possible to show that the average probability of an error after optimal addition will be defined by the formula

$$P_M = \frac{1}{2}\int\limits_0^\infty \cdots \int\limits_0^\infty \exp\left(-\sum_{i=1}^M \gamma_i\right) W_M(\gamma_1\cdots\gamma_M)d\gamma_1\cdots d\gamma_M, \tag{26}$$

where $W_M(\gamma_1\cdots\gamma_M)$ - is a combined M-dimensional density of probabilities.

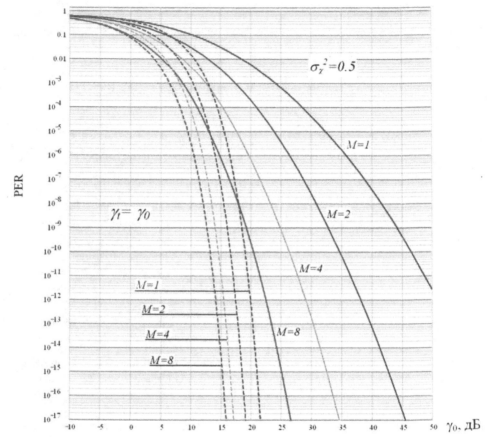

Fig. 5. Probabilities of errors of diversity reception of signals in the channel with a lognormal fading at combination of branches of diversity on algorithm of an auto select (continuous curves) and probabilities of errors at complexation of intermittent communication and diversity reception at combination of branches of diversity on algorithm of an auto select for $\gamma_i = \gamma_0$, $\sigma_\chi^2 = 0.5$ and $M=1, 2, 4, 8$.

Using the fact that the independent fading diversity branches in the combined probability density is the product of probability densities in separate branches of reception according to the analytical expression

$$P_M = \frac{1}{2}\prod_{i=1}^M \int\limits_0^\infty \exp(-\gamma_i) W_i(\gamma_i)d\gamma_i, \tag{27}$$

and with homogeneous channels in branches of diversity the probability of error reception on an output of the demodulator signal D-BPSK after operation of optimal addition will be described by the formula

$$P_M = \frac{1}{2}\left[\int_0^\infty \exp(-\gamma)W(\gamma)d\gamma\right]^M. \tag{28}$$

Formulas (27, 28) are valid at any distribution laws.

The average probability of error at complexation of the intermittent communication which are carried out in branches of diversity with the subsequent optimal addition and incoherent demodulation of signal D-BPSK at a fading of an envelope of a signal under laws Rayleigh and Nakagami will be defined accordingly by expressions (29, 30)

$$P_M(\gamma_0) = \frac{1}{2}\left[\frac{1}{\eta(\gamma_0)}\int_{\gamma_t\eta(\gamma_0)}^\infty \exp(-\alpha\gamma)\cdot\frac{1}{\gamma_0}\cdot\exp\left(-\frac{\gamma}{\gamma_0}\right)d\gamma\right]^M =$$

$$= \frac{1}{2}\left[\frac{1}{1+\gamma_0\eta(\gamma_0)}\right]^M \exp\left[-M\gamma_t\eta(\gamma_0)\right] \tag{29}$$

$$P_M(\gamma_0) = \frac{1}{2}\left\{\frac{1}{\eta(\gamma_0)}\frac{1}{\Gamma(m)}\left[\frac{m}{\gamma_0\cdot\eta(\gamma_0)}\right]^m \int_{\gamma_t\eta(\gamma_0)}^\infty \gamma^{m-1}\exp\left[-\left(\frac{m}{\gamma_0\cdot\eta(\gamma_0)}+\alpha\right)\gamma\right]d\gamma\right\}^M =$$

$$= \frac{1}{2}\left\{\frac{1}{\Gamma\left(m,m\frac{\gamma_t}{\gamma_0}\right)}\frac{\Gamma\left[m,m\frac{\gamma_t}{\gamma_0}+\alpha\gamma_t\eta(\gamma_0)\right]}{\left[1+\alpha\frac{\gamma_0}{m}\eta(\gamma_0)\right]^m}\right\} \tag{30}$$

where $\eta(\gamma_0)$ – the use factor of radio-line in a separate branch of diversity, is defined for the specified types of a fading accordingly expressions (6, 9), thus level of a threshold of interruption in various branches of diversity is identical. Analytical expression (29) is a special case of the formula (30) at m equal 1.

The probability of presence of a signal in diversity branches, under condition of their homogeneity and independence, is defined binomial by the distribution law, therefore the use factor of radio-line on an output of the circuit of optimal addition will be defined by expression (31)

$$\eta_M(\gamma_0) = 1 - \left[1 - \eta(\gamma_0)\right]^M. \tag{31}$$

The diagrams illustrating dependence of probability of an error from mean value SNR at various number of branches of diversity, according to analytical expression (30), at the complexation of intermittent communication, considered above, and diversity reception for a fading of an envelope of a signal under the law of Nakagami, at $\gamma_t = \gamma_0$ and m equal 0.6 are presented on fig. 6.

At application of intermittent communication after performance of operation of optimal addition it is more convenient to define probability density SNR on an output of the circuit of optimal addition by a method of characteristic functions, and then on the obtained probability density to calculate use factor of radio-line and probability of error.

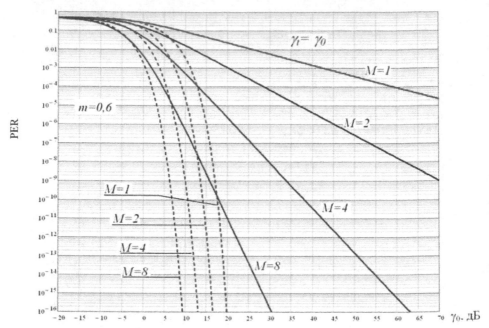

Fig. 6. Probabilities of errors at diversity reception (continuous curves) and probabilities of errors at complexation of intermittent communication carried out in branches of diversity with their subsequent optimal addition, at M equal 1,2,4 and 8, m equal 0,6 and $\gamma_t = \gamma_0$.

Characteristic function from probability density SNR at a fading an envelope of a signal under the law of Nakagami will be defined by expression

$$S(v) = \int_0^\infty f(\gamma)\exp(iv\gamma)d\gamma = \frac{m^m}{\Gamma(m)}\frac{1}{\gamma_0{}^m}\int_0^\infty \gamma^{m-1}\exp\left[-\left(\frac{m}{\gamma_0}-iv\right)\gamma\right]d\gamma .$$

(32)

Integral in the formula (32) tabular, according to expression

$$\int_0^\infty x^{v-1}\exp(-\mu x)dx = \frac{\Gamma(v)}{\mu^v} ,$$

(33)

therefore characteristic function will be defined by the formula

$$S(v) = \frac{m^m}{\gamma_0{}^m}\frac{1}{\left(\dfrac{m}{\gamma_0}-iv\right)^m} .$$

(34)

Probability density SNR in the channel with diversity reception at combination of independent homogeneous channels on algorithm of optimal addition will be defined by inverse transformation from characteristic function in raise to power of number branches of diversity

$$f_M(\gamma) = \int_{-\infty}^{\infty} \frac{1}{2\pi} [S(v)]^M \exp(-iv\gamma) dv = \frac{1}{2\pi} \frac{m^{mM}}{\gamma_0{}^m} \int_{-\infty}^{\infty} \frac{\exp(-iv\gamma)}{\left(\dfrac{m}{\gamma_0} - iv\right)^{mM}} dv . \qquad (35)$$

Integral in the formula (35) tabular, defined by to expression

$$\int_{-\infty}^{\infty} (\beta - ix)^{-v} \exp(-ipx) dx = \frac{2\pi \cdot p^{v-1} \exp(-\beta p)}{\Gamma(v)}, \qquad (36)$$

therefore final expression of probability density SNR at optimal addition branches of diversity will be defined as

$$f_M(\gamma) = \frac{m^{mM}}{\Gamma(mM)} \frac{\gamma^{mM-1}}{\gamma_0{}^{mM}} \exp\left(-\frac{\gamma}{\gamma_0} m\right). \qquad (37)$$

On fig. 7 probability density SNR is presented at optimal addition of branches of diversity, at m equal 0.6; 1.4 and 7 in the absence of the diversity, doubled and quadrupled diversity receptions.

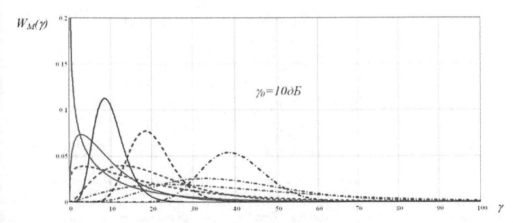

Fig. 7. Probability densities SNR at optimal addition of branches of diversity, at m equal 0.6; 1.4 and 7, accordingly red, dark blue and brown curves, in the absence of diversity; doubled and quadrupled diversity receptions, accordingly continuous, dashed and dash-dotted curves.

Use factor of radio-line, at complexation of diversity reception with optimal addition of branches of diversity and the intermittent communication which are carried out after that addition, in channels with a fading under laws Rayleigh and Nakagami are described accordingly by formulas (38, 39)

$$\eta_M(\gamma_0) = \left(\frac{1}{\gamma_0}\right)^M \frac{1}{(M-1)!} \int_{\gamma_t}^{\infty} \gamma^{M-1} \exp\left(-\frac{\gamma}{\gamma_0}\right) = \frac{\Gamma\left(M, \frac{\gamma_t}{\gamma_0}\right)}{\Gamma(M)}, \qquad (38)$$

$$\eta_M(\gamma_0) = \left(\frac{m}{\gamma_0}\right)^{mM} \frac{1}{\Gamma(mM)} \int_{\gamma_t}^{\infty} \gamma^{mM-1} \exp\left(-\frac{\gamma}{\gamma_0}m\right) = \frac{\Gamma\left(mM, \frac{\gamma_t}{\gamma_0}m\right)}{\Gamma(mM)}. \qquad (39)$$

Integrals in expressions (38) and (39) are tabular (5). Analytical expression (38) is a special case of the formula (39) at m, equal 1.

Dependences probability of error at incoherent reception of a signal at complexation of diversity reception with optimal addition of branches of diversity and the intermittent communication, which are carried out after that addition, from mean value SNR (γ_0), under formulas (40, 41) for cases of a fading of signal D-BPSK under accordingly laws Rayleigh and Nakagami will be defined by averaging of probability of errors in Gaussian noise according to the specified fading at values SNR above the given threshold level (γ_t)

$$P_M(\gamma_0) = \frac{1}{2\eta_M(\gamma_0)} \frac{1}{\gamma_0^M} \frac{1}{(M-1)!} \int_{\gamma_t}^{\infty} \gamma^{M-1} \exp\left[-\left(\frac{1}{\gamma_0}+\alpha\right)\gamma\right] = \frac{1}{2}\left(\frac{1}{1+\alpha\gamma_0}\right)^M \frac{\Gamma\left(M, \frac{\gamma_t}{\gamma_0}+\alpha\gamma_t\right)}{\Gamma\left(M, \frac{\gamma_t}{\gamma_0}\right)}, \qquad (40)$$

$$P_M(\gamma_0) = \frac{1}{2\eta_M(\gamma_0)} \left[\frac{m}{\gamma_0\eta_M(\gamma_0)}\right]^{mM} \frac{1}{\Gamma(mM)} \int_{\gamma_t\eta(\gamma_0)}^{\infty} \gamma^{mM-1} \exp\left\{-\left[\frac{m}{\gamma_0\eta(\gamma_0)}+\alpha\right]\gamma\right\} =$$

$$= \frac{1}{2}\left[1+\alpha\frac{\gamma_0}{m}\frac{\Gamma\left(mM, \frac{\gamma_t}{\gamma_0}m\right)}{\Gamma(mM)}\right]^{-mM} \frac{\Gamma\left[mM, m\frac{\gamma_t}{\gamma_0}+\alpha\gamma_t\frac{\Gamma\left(mM, \frac{\gamma_t}{\gamma_0}m\right)}{\Gamma(mM)}\right]}{\Gamma\left(mM, \frac{\gamma_t}{\gamma_0}m\right)}. \qquad (41)$$

The diagrams illustrating dependence of probability of error from mean value SNR at complexation of diversity reception with optimal addition of branches of diversity and intermittent communication, carried out after that additions, for a fading of an envelope of a signal under the law of Nakagami, at $\gamma_t = \gamma_0$ and m equal 0.6 is presented on fig. 8.

Dependences use factor of radio-line from number branches of diversity (M) at complexation of diversity reception with optimal addition branches of diversity and intermittent communication, which are carried out after that addition, and use factor of radio-line at complexation of intermittent communication, which carried out in branches of diversity with their subsequent optimal addition, is presented on fig. 9.

The use factor of radio-line at complexation of diversity reception with optimal addition of branches of diversity and the intermittent communication which are carried out after that addition, increases from magnification of branches of diversity (M) faster, than use factor of radio-line at communication of diversity reception and the intermittent communication which

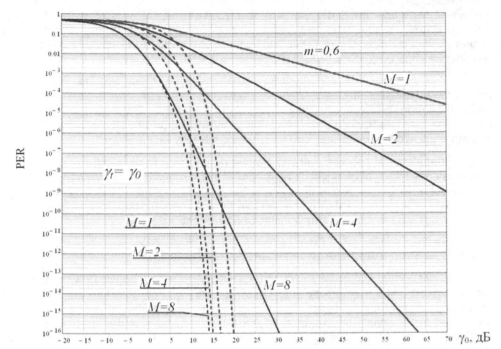

Fig. 8. Probabilities of errors at diversity reception (continuous curves) and probabilities of errors at complexation of diversity reception with optimal addition branches of diversity and the intermittent communication which are carried out after that addition, at M equal 1,2,4 and 8, m equal 0.6 and $\gamma_t = \gamma_0$.

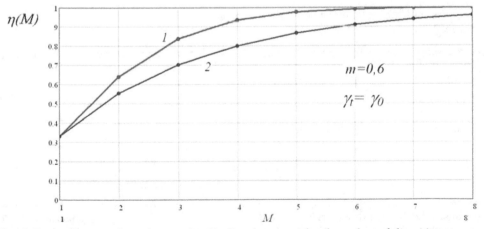

Fig. 9. Dependences of use factor of radio-line from number branches of diversity at m equal 0.6 and $\gamma_t = \gamma_0$, at complexation of intermittent communication and diversity reception when interruption is carried out on an output of the circuit of optimal addition (a line 1) and when interruption is carried out in diversity branches, with their subsequent optimal addition (a line 2).

are carried out in branches of diversity with their subsequent optimal addition (fig. 9). Values last are commensurable with use factor of radio-line at complexation of intermittent communication and diversity reception and combine of branches of diversity on algorithm of an auto select (fig. 1), even at smaller value m. Accordingly, spectral efficiency thus increases.

Probabilities of errors at complexation of diversity reception and the intermittent communication, which are carried out in branches of diversity with their subsequent optimal addition, decrease with magnification of mean value SNR faster probabilities of errors at complexation of diversity reception with optimal addition branches of diversity and intermittent communication, which are carried out after that addition.

Unlike complexation of intermittent communication and diversity reception with combination branches of diversity on algorithm of an auto select when the receiver selects a peak signal from all branches of diversity, at optimal addition the receiver should accept a signal from all branches for the subsequent handling. For this purpose it is necessary to use some antennas and to provide orthogonality of transmittable signals.

4. Complexation some methods adaptation of radio-lines

Let's consider a combination of an intermittent of radiation with constant average power and transmitter's speed automatic adjustment. We will suppose that information transfer rate changes depending on signal level in such a manner that SNR on a receiving device input remains constant. In this case at signal level changes under the Rayleigh law distribution density of probability of the random variable which is transmission rate will be defined by expression

$$W(\omega) = \frac{r}{\omega_0} \exp\left(-\frac{\omega r}{\omega_0} \right),$$
(42)

where ω is information transfer rate, ω_0 is a constant for the given system value, r – fixed value SNR

Average rate of information transfer is defined as mathematical expectation of value ω

$$V_{cp} = \int_0^\infty \omega \frac{r}{\omega_0} \exp\left(-\frac{\omega r}{\omega_0} \right) d\omega = \frac{\omega_0}{r}.$$
(43)

Median value of speed will be equal

$$V_M = \ln 2 \frac{\omega_0}{r}.$$
(44)

Let's enter designation ω_t meaning threshold value of the instant transmission rate at which the transmitter is shut down. We will find a number of the elementary signals which are transferred at speed exceeding threshold value. For this purpose we will consider integral

$$N = \int_{\omega_t}^\infty \omega \frac{r}{\omega_0} \exp\left(-\frac{\omega r}{\omega_0} \right) d\omega = \frac{\omega_0}{r} \exp\left(-\frac{\omega_t r}{\omega_0} \right) \left(\frac{r}{\omega_0} \omega_t + 1 \right).$$
(45)

Having divided it into value of average rate from (43) we will obtain a share of characters transferred at excess of threshold value ω_t

$$\frac{N}{V_{cp}} = \exp\left(-\frac{\omega_t r}{\omega_0}\right)\left(\frac{r}{\omega_0}\omega_t + 1\right). \tag{46}$$

To transfer all characters which would be transferred without transmitter shut-down it is necessary to increase average rate which we will find, having divided (43) on (46)

$$V_{cp}^{/} = \frac{\omega_0}{r}\frac{\omega_0}{r\omega_t + \omega_0}\exp\frac{\omega_t r}{\omega_0}. \tag{47}$$

The use factor of radio-line in this case is equal

$$\eta(\omega_t) = \exp\left(-\frac{\omega_t r}{\omega_0}\right). \tag{48}$$

From ratios (46) and (48) follows, that at the same level r, speed can be increased in comparison with a transmitter's speed automatic adjustment by value

$$\Delta(\omega_t) = 1 + \frac{r\omega_t}{\omega_0} = 1 + \frac{\omega_t}{V_{cp}}. \tag{49}$$

Possible gain at various values of threshold level of speed are presented in table 2

Threshold value of speed	ω_0/r	$\ln 2\omega_0/r$	$2\omega_0/r$	$3\omega_0/r$	$4\omega_0/r$
Power gain	2	$\ln 2+1$	3	4	5

Table 2.

5. Conclusion

The carried out analysis shows that in channels with random parameters (channels of systems of mobile communication, channels of tropospheric and ionospheric radio propagation, etc.) when the admissible time delay of signals exceeds duration of time intervals in which the relation a signal/noise becomes inadmissible small, application of principles of intermittent communication allows to obtain the considerable gain in noise immunity. The magnification of gain is connected to lowering of spectral efficiency and magnification of delay period of signals.

For saving of average rate of information transfer it is necessary to increase the instant speed and by that to expand occupied frequency band under a condition when in the modern communication systems basic is the requirement of saving of the frequency resource.

The analysis usage complexation of intermittent communication and diversity reception allows not only to lower probability of error an accepted signal, but also sharply to raise use factor of radio-line (spectral efficiency), especially at complexation of diversity reception with the intermittent communication which is carried out on an output of the circuit of optimal addition. The magnification of spectral efficiency is very significant owing to limitation of a resource of an electromagnetic spectrum.

The analytical ratios obtained in the given chapter, show, what gain can be implemented at usage methods of intermittent communication in various situations at the various statistics fading amplitudes of signals. These ratios allow calculating also probabilities of error at a fading of signals under laws of Rayleigh, Rice (generalized Rayleigh), Nakagami, and also lognormal laws at incoherent and coherent receptions.

In cases of possible applications for the mode of intermittent radiation (i.e., turn off the transmitter at low signal levels), the most complete communication channel can be used in a combination of variable speed and intermittent modes of communication. Potential possibilities usage methods of intermittent communication especially in a combination to other known methods of increase of noise immunity are considerable.

One of main limitation of method of intermittent communication is necessity of the considerable time delays of signals. In connection with electronics achievements this lack has the lesser value. Mastering of more and more high-frequency ranges of radio-waves leads to possibility to reduce delay period in equipment of these ranges (speed of a fading above). In connection with the marked circumstances it is possible to assume that interest to usage of methods of intermittent communication will increase.

By this time research of systems from intermittent communication is led insufficiently full. The authors suggest investigating these systems under the influence of fading interference from other radiation sources (for example, in the case Rayleigh fading as a useful signal and noise). Researches influence correlations of a fading signal in reception branches are necessary. As different variants of complex application of intermittent communication and diversity of signals (circuits MIMO) are possible, it is supposed to conduct similar researches for other variants of combined usage of intermittent communication and diversity of signals, for example at usage of different levels of thresholds at interruption's realization in different branches of diversity. Usage of intermittent communication when the signal envelope suffers simultaneously some types of a fading is of interest. For example, at propagation of signal in millimeter range through troposphere, a fading of its envelope is carried out simultaneously under laws lognormal and Nakagami. Probability density fading envelope of signal as the special case of a fading discussed (when the signal envelope fluctuates simultaneously under laws Rayleigh and lognormal and under laws Rice and lognormal) has been obtained accordingly Suzuki and Pätzold in 1977 and 1998.

6. References

Abramowitz M., Stegun A. (1964). *Handbook of mathematical functions* with Formulars, Graphs, and Mathematical Tables, National Bureau of standards applied mathematics series-55, Issued June 1964, ISBN 0-486-61272-4

Andrianov M.N., Bumagin A.V., Kalashnikov K.S. & Sisoev I. Yu. (2011) *Increase the noise immunity of data transmission over digital communication channels under Nakagami fading conditions*, Measurement Techniques, Vol.54, No. 3, 2011, pp. 57–62, ISSN 1573-8906

Andrianov M.N., Bumagin A.V., Kalashnikov K.S., Kiselev I.G. & Sisoev I. Yu. (2011) *High performance suboptimal method of transmission and reception of signals in channels with Nakagami fading*, 13th International Conference "Digital Signal Processing and its applications", Moscow, 2011, pp. 14-17, (in Russian) ISSBN 978-5-905278-01-3

Andrianov M.N., Bumagin A.V., Kiselev I.G., Rutkevich A.V., Steshenko V.B., Shishkin G.V. (2010) *The analysis and synthesis of discontinuous diversity communication for increase of efficiency of transmission of the discrete messages*, 12th International Conference "Digital Signal Processing and its applications", Moscow, 2010, pp. 82-85, (in Russian) ISSBN 978-5-904602-07-9

Andrianov M.N., Bumagin A.V., Gondar A.V., Kalashnikov K.S., Prudnikov A.A. & Steshenko V.B. (2010). *A method of increase the noise immunity and spectral efficiency of data communication channels under condition of generalized Rayleigh fading*, Measurement Techniques, Vol.53, No. 8, 2010, pp. 61–65, ISSN 1573-8906

Andrianov M.N. (2009). *Features interrupted communication of data transmission through channel under condition of lognormal fading*, Metrology, No.5, 2009, pp.35-43, (in Russian), ISSN 0132-4713.

Andrianov M.N. (2009). *Rise noise-immunity of mobile communication in lognormal fading by interconnecting interrupt communication with diversity receives*, Electromagnetic waves and electronic systems, Vol. 14, No. 8, 2009 (in Russia), pp.11-17 ISSN1560-4128

Andrianov M.N. and Kiselev I.G. (2008). *About increase of interference immunity of transmission of the discrete messages in channels with a fading*, 10th International Conference "Digital Signal Processing and its applications", Moscow, 2008, pp. 82-85, (in Russian)

Andrianov M. N. and Kiselev I.G. (2007). *Probability of error in channels with random parameters when complexing intermittent transmission with diversity reception*, Mobile. System, No. 5, 2007, pp. 44-47 (in Russian) ISSN 1729-438X

Andronov I.S. and Fink L.M. (1971). *Transmission of the discrete messages on parallel channels*, Moscow, Soviet radio, 1971 (in Russian)

Asad Mehmood & Abbas Mohammed (2010). *Characterisation and Channel Modelling for Satellite Communication Systems*, InTech, pp. 133-151, ISBN 978-953-307-135-0 , Rijeka, Croatia

Banket V. A. and Dorofeev V. M. (1988). *Digital Methods in Satellite Communications*, Radio i Svyaz, Moscow (1988) (in Russian) ISBN 5-256-00075-6

Gradshein I.S., Rizhik I.M. (1963). *Tables of Integral*, Moscow, The main publishing house of the physical and mathematical literature, 1963 (in Russian), ISBN 0-12-294757-6

Kiselev I.G. and Ustyantzev S. V. (2009) *Probability of an error in Rayleigh noisy channel at usage of interrupted signal transmission*, Mobile Telecommunications, No 1, 2009, pp. 34-36 (in Russian), ISSN 1562-4293

Kiselev I.G. and Andrianov M. N. (2007). *A method of increasing the noise protection of digital data transmission in channels with fading*, Mobile System, No. 4, 2007, pp. 13-16 (in Russian) ISSN 1729-438X

Klovsky D.D. (1976). *Processing of time-space signals (in information transfer channels)*, Moscow, Svyas, 1976 (in Russian)

Lee W. (1982) *Mobile communication engineering*, Mc Graw-Hill, 1982, ISBN 0070370397

Proakis J.G. (1995). *Digital Communication*, Mc Graw-Hill, ISBN 9007-051726-6.

Rytov S.M., Kravtsov J.A. and Tatarskiy V.I. (1978). *Introduction in statistical radio physics*, Part II. Random fields, Moscow, Science, 1978 (in Russian)

Shakhtarin B. I., Andrianov M. N., & Andrianov I. M. (2009). *Application of Discontinuous Communications in Channels with Random Parameters to Transmit Narrowband Signals and Signals with Orthogonal Frequency-Division Multiplexing*, Journal of Communications Technology and Electronics, 2009, Vol. 54, No. 10, pp. 1173–1180, ISSN 1064-2269

Shloma A.M., Bakulin M.G., Kreidelin V.B., Shumov A.P. (2008) *New algorithms of formation and handling signals in mobile communication systems*, Moscow, Gorachaya Liniya-Telecom, 2008 ISBN 978-5-9912-0061-5 (in Russian)

Sklar B. (2001). *Digital Communication, Fundamental and Application*, Second Edition, Prentice Hall PTR, ISBN 13-084788-7

Zuko A.G. (1963). *Noise immunity and efficiency system of communications*, Moscow, Svyas, 1963 (in Russian)

9

WiMAX Core Network

ZeHua Gao, Feng Gao, Bing Zhang and ZhenYu Wu

Beijing University of Posts and Telecommunications
China

1. Introduction

1.1 WiMAX core network system architecture

In the last few years, worldwide interoperability for microwave access (WiMAX) has been proposed as a promising wireless communication technology due to the fact that it can provide high data rate communications in metropolitan area networks (MANs). Until now, a number of specifications for WiMAX were standardized by the IEEE 802.16 working group. In this chapter, the architecture of WiMAX core network, all-IP mobile network, and optical integrative switching are introduced. It also tells the accounting of WiMAX in detail.

1.2 Mobile network core network architecture

WiMAX network structure mainly includes WiMAX Access Service Network and WiMAX Connectivity Service Network. CSN is the core network of WiMAX. It provides the IP-connection service for the users. Its connection modules are shown in the figure below:

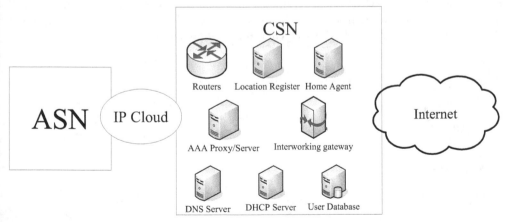

Fig. 1. CSN connection modules

As you can see in CSN, it includes the router, location register, home agent and AAA-server. Router connects CSN with the other modules. Location register record the user's login and location information. In order to support the mobility, CSN provides mobile IP function.

Home agent is responsible for maintaining MS position imformation and sending the packets to the network of MS.AAA proxy/server provide authentication, authorization and accounting services. As to connect Internet or any other IP network, CSN may also include user database and interworking gateway devices, DHCP server and DNS server.

CSN is defined as the combination of network function. It includes these performances:

a. MS IP address and endpoint parameter allocation for user sessions
b. Internet access
c. AAA services
d. Policy and Admission Control based on user subscription profiles
e. ASN-CSN tunneling support
f. WiMAX subscriber billing and inter-operator settlement
g. Inter-CSN tunneling for roaming
h. Inter-ASN mobility
i. Connectivity to WiMAX services such as IP multimedia services (IMS), location based services, peer-to-peer services and provisioning.

ASN and CSN are not belong to the same operator. CSN can use several ASNs and it is the same to ASN. Several CSNs can share the services that just one ASN provides. In this case, it will change information between ASN and MS to let ASN know that which CSN is connected with MS.

If WiMAX is arranged alone, CSN can be used as independent network construction. If it constructed with the other network, they can share some function entity.

1.2.1 All-IP mobile network

The mobile broadband wireless industry is in the midst of a significant transition in terms of capabilities and means of delivering multimedia IP services anytime, anywhere. So the all-IP mobile network specification is being defined by the Network Working Group (NWG) in the WiMAX Forum. The ip-based WiMAX architecture is shown in the figure below:

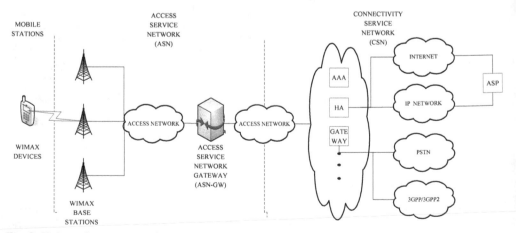

Fig. 2. IP-based WiMAX architecture

And its reference model is shown in the figure below:

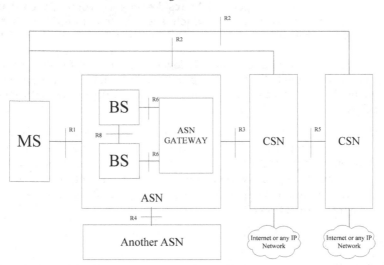

It gives the reference network model interfaces:

R1 Interface between the MS and the ASN. Functionality: air interface.

R2 Interface between the MS and the CSN. Functionality: AAP, IP host configuration, mobility management.

R3 Interface between the ASN and the CSN. Functionality: AAP, policy enforcement, mobility management.

R4 Interface between ASNs. Functionality: mobility management.

R5 Interface between CSNs. Functionality: internetworking, roaming.

R6 Interface between BS and ASN gateway. Functionality: IP tunnel management to establish and release MS connection.

R8 Interface between BSs. Functionality: handoffs.

R7 is the interface in ASN gateway. It is not shown in figure 3.

Fig. 3. IP-based WiMAX network model

A prominent feature of the NWG specification is the extensive use of IP and IETF-standard protocols. The focus is on enabling IP access for mobile devices. Networking functionality requirements for client devices consist of just standard IP protocols like DHCP, Mobile IP, EAP protocols. IP connectivity is assumed between all interacting entities in the network. Mobile IP used as the mechanism for redirection of the data as a mobile device moves from one ASN to another ASN, crossing IP subnet/prefix boundaries. Mobility support for mobiles that are not Mobile-IP capable is provided by the use of Proxy Mobile IP. On the network side, IP address pool management provided through IETF standard mechanisms like DHCP or AAA. Decomposition of protocols across reference points enables interoperability while accommodating flexible implementation choices for vendors and operators.

WiMAX network architecture utilizes a combination of the IETF-standard Mobile IP protocol and special protocols defined by WiMAX NWG to handle mobility. The use of standard Mobile IP makes it possible to leverage off-theshelf components such as Home Agent in addition to simplifying the interface to the rest of the IP world.Alongside IETF Mobile IP, WiMAX specific protocols are used to allow for optimizations and to provide flexibility in handling mobility. For a given MS, the Mobile IP Home Agent (HA) resides in a CSN and one or more Foreign Agents (FA) resides in each ASN. Data for this MS is transported through the Mobile IP tunnel, which is terminated at an FA in an ASN. Once Mobile IP tunnel is terminated at the FA, WiMAX specific protocols (i.e. Data Path Functions) take over and transport the data from the FA to the serving base station, to which the MS is attached.This design offers multiple levels of anchoring for the user data plane path during handovers. Mobile IP is used to provide a "top" level of anchoring of the data flow for a mobile. WiMAX specific Data Path Function (DPF) to provide further levels of anchoring "below" Mobile IP.

Note that many mobile clients today such as PDAs and laptops are not Mobile-IP capable; instead they work with simple IP. WiMAX network architecture assures that those clients, together with Mobile-IP capable clients, are fully able to use services offered by the network. This is achieved by employing Proxy Mobile IP Client (PMIP Client), which acts as a proxy for the mobile client, in the network and handles the Mobile IP procedures in lieu of the mobile client in a transparent fashion.

1.2.2 Optical integrative switching

With the increasing of data traffic and the growing diversity of services, several optical network paradigms for future internet backbone have been under intensive research. Wavelength division multiplexing (WDM) appears to be the solution of choice for providing a faster networking infrastructure that can meet the explosive growth of the Internet. Since this growth is mainly fueled by IP data traffic, wavelength-routed optical networks employing circuit switching which is also called optical circuit switching (OCS) may not be the most appropriate for the emerging optical Internet. OCS is lack of flexibility to cope with the fluctuating traffic and the changing link status. Optical packet switching (OPS) is an alternative technology that appears to be the optimum choice. The OPS requires the technologies such as optical buffer and optical logic which is not mature enough to provide a viable solution.

Optical burst switching (OBS) represents a balance between optical circuit switching and optical packet switching that combines the best features of both. A comparison of optical burst switching approaches shows that OBS delivers a high-bandwidth utilization, cost efficiencies and good adaptivity to congestion while avoiding OPS's implementation difficulties such as high processing and synchronization overhead and the need for too immature optical buffer memory. OBS delivers a high-bandwidth utilization and it is much agile than the OCS. While OBS is not good at transmitting circuit switching services.

Optical Integrative Switching (OIS) networks technology can agilely support the multi-services[1 IEEE Std 802.16e] in WIMAX. In OIS ring networks technology supporting WiMAX, four different classes of service traffic are considered. The 0 class of traffic (Class 0) can support Unsolicited Grant Service (type 0) in WiMAX; The class 1 traffic can support Real-Time Variable Rate Service (type 1) and Extended Real-Time Variable Rate Service (type 4) in WiMAX; The class 2 traffic can support Non-Real-Time Variable Rate service (type 2) in WiMAX; The class 3 traffic can support Best Efforts Service (type 3) in WiMAX. For service flows, the class parameter should be used when determining precedence in request service and the lower numbers indicate higher priority.

1.2.2.1 System architecture

Ring and node architecture

OIS can be mesh network or ring network. The OIS ring network is shown in figure below.

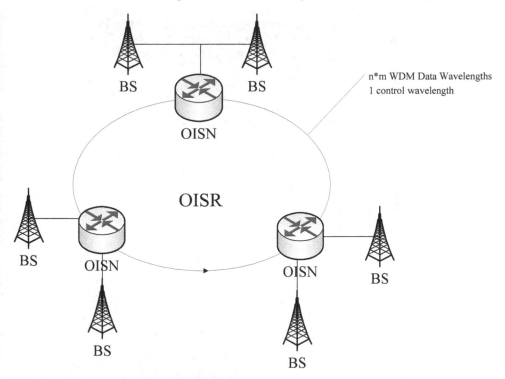

OISN: Optical integrative switching ring node
OISR: Optical integrative switching ring network
BS: WiMAX Base Station
The system has n OIS nodes in an unidirectional OIS ring. The OIS ring can be a backbone of wireless metropolitan area network (MAN) that interconnects a number of WiMAX base stations.

Fig. 4. Architecture of OIS ring network

The optical integrative switching ring network has n×m+1 wavelengths. The total data transmitters number in the OIS ring network is n×m corresponding n×m WDM wavelengths. Each node has m wavelengths which are referred as the home wavelengths of the node. The (n×m+1)th wavelength is the control channel for the wavelengths in the set and it carries control frames. The control frames implement the signaling necessary for OIS. Each OIS node in the ring has m data transmitters each fixed-tuned to one of the m home wavelengths, and m tunable receivers. These m pairs of transceivers are used for transmitting and receiving bursts. A node can only transmit bursts on its home wavelengths. A free receiver can tune to receive a burst arriving on any wavelength in the ring. A WTFM (wavelength tunable filter module) was installed at the destination node to drop any m data wavelengths and the control wavelength. The network scheme has the feature of transmitting wavelengths collision free, traffic collision free when switching at the midst nodes, traffic collision free from the source node. The core WDM network utilizes the W-token (token supporting WiMAX) signaling scheme to resolve receiver collisions at the receivers. In the W-token access protocol, the token can be captured according to the service class. The OIS node is equipped with a control module which performs its functions based on the information each control frame carries around the ring. Each node has its own slot into which it can write information during transmission. The control frames on the m control wavelengths travel around the ring in a synchronous manner.

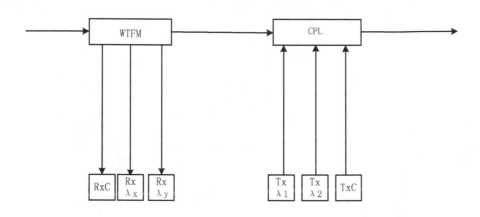

CPL: coupler
WTFM: wavelength tunable filter module
RxC: the receiver of control channel
TxC: the transmitter of control channel
Rx: the receiver of traffic channel
Tx: the transmitter of traffic channel

Fig. 5. The Node Architecture of OIS ring network when m=2

Each OIS node serves a number of WiMAX base stations. In the direction from the WiMAX base stations to the ring, the OIS node collects and buffers electronically data packets, transmitted by users over the WiMAX base stations. Buffered packets are subsequently grouped in different burst according to their classes and destination. A burst can be of any size between a minimum and maximum value. Bursts travel as optical signals along the ring, without undergoing any electro-optic conversion at intermediate nodes. In the other direction from the ring to the WiMAX base stations, an OIS node terminates optical bursts destined to itself, electronically processes the data packets contained therein, and delivers them to users in its attached WiMAX base stations.

A WTFM was installed at each destination node before m data receivers and the control channel receiver to drop any m data wavelengths and the control channel wavelength. The architecture of the node when m=2 is shown as Fig.2. When the node receives the wavelengths from the ring, the two data wavelengths and the control channel wavelength was dropped by the WTFM. The other wavelengths were coupled with the two home data wavelengths (λ1 and λ2) which transmit the burst of the node as well as the control channel wavelength and transmit to the OIS ring network.

In OIS ring networks technology supporting WiMAX, there are four different classes of service traffic. The 0 class of traffic (Class 0) can support Unsolicited Grant Service (type 0) in WiMAX; The class 1 traffic can support Real-Time Variable Rate Service (type 1) and Extended Real-Time Variable Rate Service (type 4) in WiMAX; The class 2 traffic can support Non-Real-Time Variable Rate service (type 2) in WiMAX; The class 3 traffic can support Best Efforts Service (type 3) in WiMAX. For service flows, the class parameter should be used when determining precedence in request service and the lower numbers indicate preemptive priority. Bursts with different classes are served by using W-token access protocol. In the W-token access protocol, token can be released and captured according to the service class. In the OIS network, token0 and token1 are distributed for every receiving port to indicate the receiving port is unoccupied. Token0 can support class0 service and token1 can support other service in OIS network.

Control wavelength operation

The control channel wavelength is used for the transmission of control signal. In a ring with n nodes, n×4×m control slots, 4×m slots corresponding to m transmitters of each node, are grouped together in a control frame which continuously circulates around the ring. Four control slots includes token0, token1, answer frame (ACK) and a control frame (CF) corresponding to each transmitter of each node.

Each node is the owner of 4×m control slots in each control frame. Each control slot contains several fields. Each token0 control slot includes fields for the destination address, the token serial number, class of traffic is 0, the offset, the transmitter wavelength of exist class0 service, the time slot of exist class0 service, the time slot of reserving class0 service and so on. Each token1 control slot includes fields for the destination address, the token serial number, class of traffic, the offset, the transmitter wavelength, the time slot of exist class0 service, the time slot of reserving class0 service and so on. Each answer frame (ACK) includes fields for the destination address and the source address, the token0 serial number, class of traffic, the offset, flag, the transmitter wavelength and the size of data, the time slot of exist class0 service, the time slot of reserving class0 service and so on. Each CF control slot includes fields for the destination address, the token1 serial number, class of traffic, the offset, flags, the transmitter wavelength, the burst size the time slot of exist class0 service, the time slot of reserving class0 service and so on.

The value of traffic class is as follows: 0 is corresponding to class 0 traffic, 1 is corresponding to class 1 traffic 2 is corresponding to class 2 traffic, 3 is corresponding to class 3 traffic, and 9 is corresponding to no traffic. The offset 1 value is the processing times at intermediate nodes and the offset 2 value is the time of WTFM to switch to receive different wavelength. The offset value is the sum of the offset 1 and the offset 2. The offset time doesn't include the transmitting time from source to destination.

1.2.2.2 The W-token access protocol in OIS network

The protocol uses tokens to resolve receiver collisions at the receivers. Tokens are used for all classes. Every node has m token0 and m token1 corresponding to m receivers of this node circulating around the ring. If a source wants to transmit a burst to a particular destination, it has to catch the token for that destination according to classes of service. To accommodate the extant PCM network (including SDH), 125μs is adopted as data period for receiving port. A token may be either available or in use. When the sending node catches token0, the time slot of receiving port will be reserved. If the time slot of receiving port reserved successful, the connection of class0 will be established. Then the class0 occupy the same time slot in every period of 125μs. This message of occupying time slot will be write in token. The else service will keep away from this time slot when they catch token.

Since only the node that has possession of the token can transmit a burst to the appropriate destination in its time slot, the receiver of the destination can only receive a single burst at a time, and therefore the W-token protocol is a receiver collision-free protocol.

The maximum bandwidth of each class can be defined by the network. The service bandwidth which has a lower priority can be taken up by the higher priority; however, the defined maximum bandwidth can not be exceeded. Processes of each class service sending as below:

The processes of sending class 0 service

At the sending node, the node checks each received control frame. If the node detects an available token0, it deletes token0 from control frame, and puts token0 into its own FIFO token queue. If there is no class 0 service, the node will release token0 to the next node, otherwise, it will examine the receiving port time slot pool, and detect whether there are available time slots of the receiving port which is limited by the network (The time slot which is not occupied by class 0 can be regard as an available time slot. When the time slot is occupied by lower priority service, it can be grabbed by higher ones. However, the class 0 bandwidth can not be wider than the maximum class 0 bandwidth which is defined by network. According to the network resources, network can define the maximum class 0 service bandwidth of the whole network. The maximum connections bandwidth between each pair of nodes can be defined. Idle available time slot means it is longer than data required). If there is none, the node will release token0 to the next node, else, it will examine local sending time slots, and see whether there are idle available sending time slots (If the total bandwidth including the bandwidth which is required by the sending data is not larger than the class 0 service maximum bandwidth, then it has idle sending time slots, or else it has no idle sending time slots. Idle available sending time slot means the idle time slot is larger than required time slot of data sending),if there is no idle available time slots, the node will release token0 to the next node, otherwise it will reserve sending time slots, and write reserving message into token0, then release token0 to the next node.

When the receiving node reveives token0, it will send the controlling order to WTFM to receive the wavelength which carrying data during the time slot and to receive class 0 service data according to the reserved time slots. Class 0 service coming from different source nodes is received according to the time slots in the frame with the period of 125μs. The frame period 125μs is defined for receiving ports, and different sending nodes occupy different time slots of receiving port. Because the wavelengths of different nodes are different, the WTFM may be used at the receiving ports. Only one OIS class 0 service data packet from one sending node is sent in each frame with the period of 125μs (Multiple E1 frames may be included in the period of 125μs.). Class 0 service supports WiMAX type0 service. When the receiving port was reserved to receive class0 service, the receiving node will send an ACK message to the source node and the receiving node writes arrangement message to token0 and releases it to the next node. When receiving token1 and CF, it writes into the time slots information, and releases it to the next node. After the source node receives ACK which indicates that reserving time slots is successful, it starts to send class 0 service. When finished sending data and receiving token0, the source node write the releasing message into token0 and releases it to the next node. When received the token0 including the releasing message, the receiving node sends controlling order to the WTFM and releases the connection. It writes this time slot releasing message to token0 and ACK and release to the next node. When receiving token1 and CF, the time slot releasing information is updated and token1 and CF released to the next node.

The processes of sending class 1 service

At the sending node, it checks each received controlling frame. When it detects an available token1, then it deletes token1 from the controlling frame, and puts token1 into its own FIFO token queue. If there is no class1 service, the node will check whether there is class2 or class3 service. If there is class1 service, it will check whether there are idle available time slots at the receiving port (Idle time slots of the receiving port here is the time slots not occupied by class 0 or class 1. When the time slot is occupied by lower priority service, it can be grabbed by higher ones. However, the maximum class 1 bandwidth defined by network should not be exceeded. According to the network resources, the network could define the maximum class 1 service bandwidth of the whole network, and also it could define the maximum connection bandwidth between each pair of nodes. Idle available time slot means it is longer than burst packets required), if there is none, the node will check whether there is class2 or class3 service. If there are idle available time slots, then it checks whether there are local idle available sending time slots limited by the network (If the sum of the bandwidth of the existed class 1 service and the required bandwidth of sending time slots is not larger than the maximum bandwidth of class 1 service defined by the network, then the node will consider that there are idle sending time slots, otherwise, there are none. Idle available sending time slot means that it is larger than the burst packet sending time slot, and the time slot doesn't conflict with the idle time slot of the receiving port),if there are none, the node will go on to check whether there is class2 or class3 service. If there are idle available sending time slots, the message of occupying time slots will be written into token1, and then to check whether there is class 3 service. At the same time, the sending node writes the message about the occupied time slots into the control message CF and sends it to control frame. And then sends class1 data after the offset time. Every time the node captures token1, it could send multiple OIS class1 service burst packets in different frames whose period is 125µs (each class 1 service burst packet can contain multiple MPEG frames etc). Only one OIS class1 service burst packet is sent in each frame with the period of 125µs. The total bandwidth should not larger than the maximum bandwidth defined by the network. Receiving CF, the receiving node will send control order to the WTFM and receive class1 service data according to the reserved timeslot.

The processes of sending class2 service

Check whether there is class2 service, and if there is none, it will check whether there is class3 service. If there is class2 service, it will check whether there are idle available time slots in the receiving port (The idle available time slot here means the time slots which are not occupied by class0, class1 and class2. The service can grab the time slots of service with lower priority. The total bandwidth of class2 can not be larger than the maximum class2 service bandwidth defined by the network. According to the network resources, the network can define the maximum class2 service bandwidth of the whole network. It can also define the maximum connection bandwidth between each pair of nodes. Idle available time slot means it is longer than the burst packets required), if there is none, the node will go on to check whether there is class3 service. If there are idle available time slots, it will check whether there are local idle available sending time slots limited by the network (If the sum of the bandwidth of existing class2 service and the bandwidth of sending time slots required is not larger than the class2 service maximum bandwidth defined by the network, then the node will consider that there are idle sending time slots, otherwise it considers there are none. Idle available sending time slot means it is longer than that required by the burst packet, and the time slot does not conflict with the idle time slot in the receiving port), if there is none, it will go to check whether there is class3 service. If there are idle available sending time slots, it will write the message about time slot occupying into token1, then go to check if there is classs3 service. At the same time, the occupied time slot message is written into the control message CF and then CF was sent. The class2 data was sent after the offset time. When the node captures token1, it can send multiple OIS class2 service burst packets in different frames with the period of 125µs. When receiving CF, the receiving node sends control order to WTFM and then receives class2 service data according to the reserved time slots.

The processes of sending class3 service

The node checks whether there are idle available time slots at the receiving port (Idle available time slot at the receiving port is the time slot which is not occupied by class0, 1, 2 and 3. However, the class 3 bandwidth can not be wider than the maximum class 3 bandwidth which is defined by the network. According to the resources, the network can define the maximum class3 bandwidth of the whole network. The maximum connections bandwidth between each pair of nodes can be defined. Idle available time slot means it is longer than burst packets required), if there is none, token1 will be released to the next node. Otherwise, it will check whether there are local idle available time slots limited by the network. (If the sum of the bandwidth of existing class3 service and the required bandwidth of sending time slots is not larger than the maximum bandwidth of class 3 service defined by the network, then the node will consider that there are idle sending time slots, otherwise it considers there are none. Idle available sending time slot means it is longer than that required by the burst packet, and the time slot does not conflict with the idle time slot in the receiving port), if there is none, the node will release token1 to the next node. If there are idle available sending time slots, it will write the message about time slot occupied into token1, then release token1 to the next node. At the same time, the occupied time slot message is written into the control message CF and then CF was sent. The class3 data was sent after the offset time. When the node captures token1, it can send multiple OIS class3 service burst packets in different frames with the period of 125µs. When receiving CF, the receiving node sends control order to WTFM and then receives class3 service data according to the reserved time slots. After the receiving node receives token1, it will write the message about the time slots occupied by class 0 service into token1.

2. WiMAX core network main function

Considering the main function of the WiMAX core network, it includes the exchanging with the existing network and accounting. So these points below should be thought about:

1. WiMAX network architecture should be considered working together with existing wireless or wired networks such as GSM, WCDMA, Wi-Fi, DSL and so on. Meanwhile, they should be based on IP/IETF protocol standards.
2. WiMAX network architecture will also support global roaming, billing and settlement systems in public.
3. WiMAX network architecture will support a wide variety of user authentication, authentication methods, such as based on user name / password, digital certificate, SIM, USIM.

2.1 Interoperability with 3GPP networks

WiMAX network has the features with high spectrum efficiency, large data throughput, and effective cost. While 3GPP network has the mature services of voice, roaming, handover, improved authentication and accounting.Achieving the interoperability of these different networks will meet the future needs of the growing market.

There are many programs to achieve the interoperability of these two networks but operators must determine the details according to the actual situation. Generally, the interoperability between 3GPP and WiMAX has the following six application modes:

1. Unified billing and customer relationship mode: users sign with two network operations, and it is able to access to both networks, but use one list of all settlement costs
2. Based on WCDMA accessing control and billing: certification, authentication and billing are provided by WCDMA.
3. WCDMA PS services: users who access via WiMAX network can directly use WCDMA PS services, such as LCS, MMS, instant messaging, etc.
4. WCDMA & WiMAX continuous service: Users can conduct sevices switching in both of two networks, it may cause the pause and packet loss.
5. WCDMA & WiMAX seamless service: Users can conduct sevices switching in both of two networks, it can not cause the pause and packet loss.
6. WCDMA CS services: users who access via WiMAX network can directly use WCDMA CS services, and provide seamless handover.

In these modes, WiMAX can be added to or be side by side with WCDMA.

Considering current network deployment, existing 3GPP network and commercial development progress of WiMAX, it is recommended the following two interconnections:

- WiMAX is attached to the WCDMA; two networks belong to the same operator.
- WiMAX's authentication, authorization and accounting are conducted by AAA, HLR and other network elements of WCDMA.
- WiMAX transport the data through its own AC (AGW), directly connect to the external PDN. Users' accounting information is collected at AC and then report to the accounting system of WCDMA.

- WiMAX works as an additional network and cause less impact on WCDMA.

This program is suitable for the case that WCDMA construction is completed. It can enhance the data rate by deploying WiMAX.

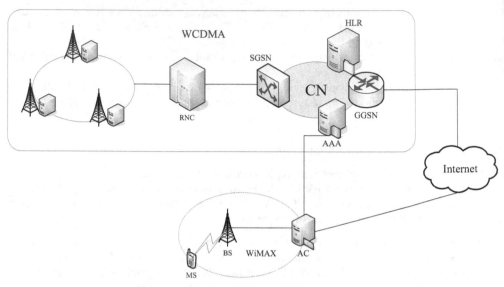

Fig. 6. First Plan for Interoperability with 3GPP networks

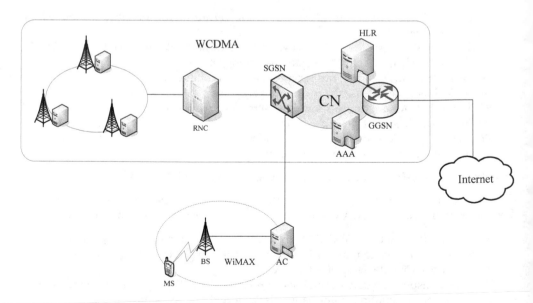

Fig. 7. Second Plan for Interoperability with 3GPP networks

- WiMAX is attached to the WCDMA . The authentication and authorization of WiMAX users are completed by WCDMA HLR.
- SGSN packages WiMAX data and sends it to GGSN. GGSN is responsible for routing it to the external PDN.
- According to the actual deployment of the integration,WiMAX users can access to PS services of WCDMA to achieve more closer integration.
- This plan involves all of WCDMA core network elements. There is a greater impact on the system and it has a higher risk.

The first plan is for initial solution of interconnection between two networks. With the development of this service, you can choose the second plan. Data services will be open to the users of WiMAX gradually.

2.2 Accounting

2.2.1 The network structure of accounting

The accounting structure of WiMAX access layer is shown in the figure below:

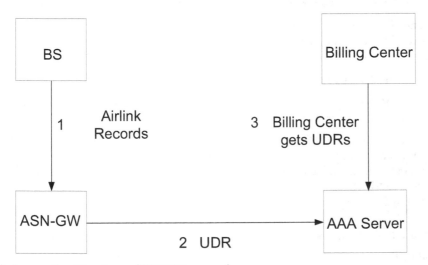

Fig. 8. Accounting Structure of WiMAX access layer

WiMAX accounting consists of two parts: BS collects the accounting parameters of the wireless side and ASN-GW collects the accounting parameters of the network side. BS sends the wireless side accounting parameters to ASN-GW through the interface between them. ASN-GW integrates the wireless side and network side accounting parameters into UDR and then sends it to AAA-server through RADIUS interface. AAA-server provides the interface to the accounting center and it can collect the initial voice information from AAA-server.

The interface between ASN-GW and AAA-server supports RADIUS protocol and the evolution to DIAMETER protocol. Considering the special bits of WiMAX accounting, it will use RADIUS or DIAMETER according to the producers.

2.2.2 Accounting modes

Fixed accounting

According to the user's contract situation, fixed accounting makes user do not need to consider the service usage when he access to WiMAX to use all sorts of services. When the user's contract expires, the system will stop user from accessing the network. Fixed accounting includes the following four modes:

- Accounting according to the conversation.
- Accounting according to the conversation and limit the maximum duration in each session. E.g. a longest session is 24 hours.
- Accounting according to the conversation and limit the period in one day to access the network. E.g. from noon to night.
- Accounting according to the time and do not consider how many services the user uses. E.g. account monthly.

Accounting based on services usage

This mode is according to the services usage of user. There are three methods:

- Duration of the conversation.The actual use of time in a continuous session.
- Data traffic. The user and air interface transport and receive data bytes
- Sessions number. The number of sessions which are successfully established.

Accounting based on services

This mode is according to all kinds of the value-added services supported by operators. There are three possible methods:

- User information contains the particular IP applications (e.g. VOIP),   QoS/SLA agreement and some kind of roaming agreement in the contract.
- Temporary and dynamic authorized content by the network (such as the conversation accessed to VPN).
- Using third party to provide the services in the conversation.

2.2.3 Accounting solutions

Post-paid billing

This mode is that user uses services first and then the operator charges according to the usage of the services.

When the user's session is set up, the system generates the accounting start list of this session. During the session, the system generates the accounting interim list at an interval of some time. When the session ends, the system generates the accounting stop list. Here is the figure of its procedure:

Its procedure is mainly about these points:

a. User establishes the session through the authorization of AAA.
b. ASN-GW send the message to AAA to start accounting.
c. After some time of this session, the system billing timer is overtime.

d.   ASN-GW sends accounting message to AAA in the middle of the session.
e.   The user ends up the session.
f.   ASN-GW send stop accounting request to AAA.
g.   Billing center uses the interface provided by AAA to collect the accounting records.

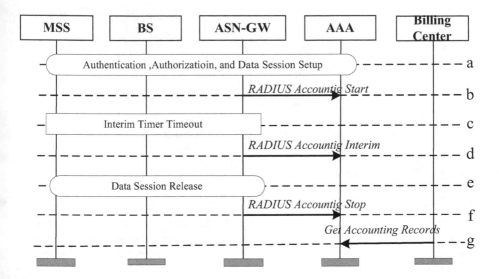

Fig. 9. Post-paid Billing Accounting Procedure

Pre-paid billing

This mode is that the user pays first and the operator check user account information. The user uses the services according to the balance in his account. Pre-paid billing can get the cost of settlement at real-time. According to the balance of the user's account, the operator can control the user's usage of services to ensure the interest of itself.

The difference between pre-paid billing and post-paid billing is that pre-paid billing must pay first and then use the services. The user's accounting information is saved in the system previously. ASN-GW get the information of the balance and during the services ASN-GW diminish this balance.When the balance runs out, ASN-GW request the system to redistribute the account balance, and so on in turn. When all of the user's accounting balance runs out, the system will terminate user's services initiatively.

At present WiMAX can not support pre-paid billing. You can see the following exsiting reference structure of pre-paid billing:

Pre-paid billing entity includes:

AAA-server: Provide authentication, authorization and accounting services.

PPS: Pre-paid server. Control the services of pre-paid billing user.

PPC: Pre-paid client. Provide services according to the PPS pre-paid control strategy.

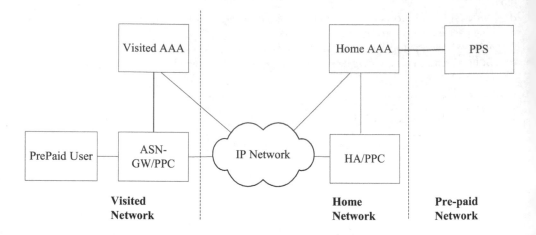

Fig. 10. Pre-paid Billing Reference Structure

Roaming billing

In the actual roaming environment, it may include many scenarios and network configuration. In order to manage this complexity, it defines an universal roaming model and it is shown in the following figure. There is only one accounting mode that the operator send the bill to users.

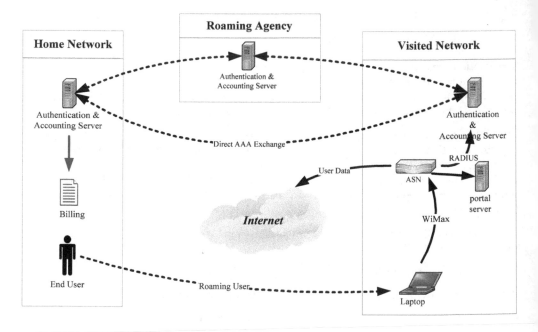

Fig. 11. Universal Roaming Model

In the above roaming model, WiMAX is divided into three parts: Visited Network, Roaming Agency and Home Network. By exchanging the information between these parts, it can provide authentication, authorization and accounting services to ensure the user to access to WiMAX.

In WiMAX, roaming agency is an optional part. It will use roaming agency or not according to the actual need. Roaming agency takes on an intermediary role in multiple roaming. In the actual roaming environment, an operator must communicate with several operators to realize the roaming of WiMAX. If they communicate with each other directly, it will be very difficult and its cost will be very large. However, if a operator access to the roaming agency, it can communicate with the other operators which are attached to the roaming agency. This is similar to the structure of Internet. Currently iPass and GRIC are most influential in roaming agency.

Roaming agency can be divided into two parts:

a.   Route based on the authentication information. It sends the authentication information from visited network to home network to realize the user's authentication by proxy function of radius protocol. In this process, the authentication information may experience several proxies, which depends on the network environment.
b.   Accounting. Some roaming agencies undertake the function of billing center to deal with accounting in roaming.

3. Concluding remarks

In summary, it can be observed that the future network will have the following characteristics: IP-based, broadband, lower cost, more convenient to manage, and multiple technologies could integrate through a unified core network, achieving the the seamless switching between various services.

The core network of WiMAX adopts mobile IP framework, which provides the ability of seamless integration with all-IP network. It has following advantages:

1.   WiMAX core network meets the QoS requirements of various service and applications, effectively utilizing end-to-end network resources.
2.   The extendibility, flexibility, and robustness of WiMAX core network help to reach the demand for telecommunication level network deployment.
3.   WiMAX core network is equipped with advanced mobility management, including paging, location management, cutover between various technologies, and cutover between different operator networks.
4.   WiMAX core network could provide the security and QoS indemnification in all mobile patterns.

In the next stage, WiMAX core network will gradually evolve towards IMS, adapting to the requirements of interconnecting with other network.

4. References

[1] WiMAX Forum, NWG, WiMAX End to End Network System Architecture(Stage2: Architecture Tenets, Reference Model and Rerefence Points), 2006

[2] 3GPP TS 24.402 3GPP System Architecture Evolution: Architecture Enhancements for non - 3GPP accesses, Release 8, 2007

[3] XiaoLu Dong, MeiMei Dang, WiMAX technology, standards and applications. Beijing : Posts and Telecom Press, 2007

[4] 3GPP TS 24.008 Mobile radio interface Layer 3 specification, Core network protocols, Stage 3 (Release 7), 2006

[5] XiongYan Tang, Broadband wireless access technologies and applications - WiMAX and Wi-Fi. Beijing: Publishing House of Electronics Industry, 2006

10

The WiMAX PHY Layer

Marcel O. Odhiambo and Amimo P.O. Rayolla

University of South Africa
Department of Electrical and Mining Engineering, Florida
South Africa

1. Introduction

This chapter will explore The PHY layers defined in IEEE 802.16d&e (2004/5 standards updates), OFDM/OFDMA and SOFDMA in the frequency domain, Symbol mapping and channel encoding, Link Adaptation - Channel coding and modulation schemes, Control mechanisms and antenna diversity and spatial multiplexing.

The discussions thereof shall thus be based on IEEE 802.16d, herewith referred to as *fixed WiMAX* and IEEE 802.16e herein referred to as *mobile WiMAX*. The specific PHY technical descriptions are given in the WiMAX Forum-T21/3 documentations or the original IEEE 802.16d suite of standards [2, 3, 4].

The PHYsical Layer in Fig. 1 provides the air interface between the Base Station (BS) and the Subscriber Stations (SS/MS) in different frequency bands for the entire range of IEEE 802.16* standards for single and multi-carrier bands with OFDM/OFDMA and SOFDM/SOFDMA. PHY) layer thus takes MAC PDUs at the PHY SAP and arranges them for transport over the air interface [4, 5].

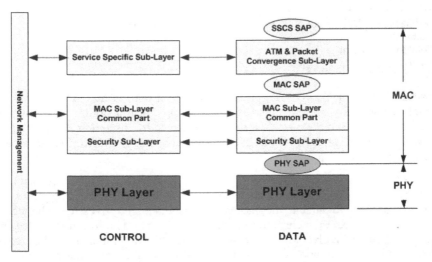

Fig. 1. WirelessMAN Protocol Layers

IEEE 802.16 (2001) specified PHY in 10-66 GHz range. This was further updated in 2004 and 2005 to 802.16d and 802.16e to define the 2-11 GHz range with enhancements such as Scalable Orthogonal Frequency-Division Multiple Access (SOFDMA) to the original multiplexing technique for fixed WiMAX, Orthogonal Frequency-Division Multiplexing (OFDM). Further updates have since been done in 2009 and 2011 to define Fixed and Mobile Broadband, and Mobile WiMAX respectively. In between these updates are several mergers, superseded and withdrawn projects. This chapter will NOT give further discussions on 2009 and 2011 updates to the 802.16 standard which have introduced multi-hop relay features (2009) and enhanced mobility and data rate features (2011).

The IEEE802.16 suite of standards defines four PHY layers in the Licensed Band namely WirelessMAN SC, WirelessMAN SCa, WirelessMAN OFDM (IEEE 802.16-2004) and WirelessMAN OFDMA (IEEE 802.16-2004), with further modifications to a Scalable OFDMA with a further one in the Unlicensed Band, WirelessHUMAN as summarized in Table 1 [2, 3].

PHY	Propagation	Operation	Freq Band	Carrier	Duplexing
WirelessMAN SC	LOS	P2P	10-66 GHz	Single	T/FDD
WirelessMAN SCa	LOS	P2P	2-11 GHz	Single	T/FDD
WirelessMAN 16d (OFDM/A)	NLOS	PMP	2-11 GHz	256	T/FDD
WirelessMAN 16e (S/OFDMA)	NLOS	PMP	2-11 GHz	2048	T/FDD
WirelessHUMAN*	NLOS	PMP	2-11 GHz*	1/256/2048	TDD, Dynamic Frequency Selection

* is for the High-speed Unlicensed band using license except frequencies in the 2-11 GHz band.
P2P is Point to Point
PMP is Point to Multi-Point

Table 1. Brief of IEEE 802.16 PHY

A functional WiMAX PHY Layer is represented by Fig. 2 both in time-frequency domains and digital-analog domains. The figure is shown from the transmitter end which is by default what's defined and the receiver is mostly left to vendor discretion. The first stage has to deal with Forward Error Check (FEC), channel encoding, puncturing or repeating, interleaving, and symbol mapping.

The next functional stage is the construction of the OFDM symbol in the frequency domain by mapping data onto the appropriate sub-channels/subcarriers and inserting pilot symbols into pilot subcarriers to enable the receiver to estimate and track the channel state information (CSI). Space/time encoding for transmit diversity is also implemented in this stage. The final stage involves the conversion of the OFDM symbol from the frequency domain to the time domain and eventually to an analog signal for transmission over the air interface [9, 4].

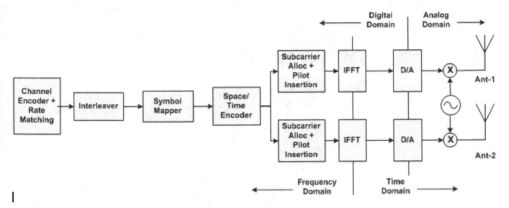

Fig. 2. WiMAX PHY Functional Diagram in Time-Frequency & Digital-Analog Domains

2. OFDM

Orthogonal Frequency-Division Multiplexing is multicarrier modulation technique with dedicated carrier spacing in the frequency domain. The orthogonality aims to address the Inter-Symbol Interference (ISI) while bringing better data rates with increased numbers of sub-carriers. If the modulated carrier is represented by a $sin\,x$ function, then the sub-channels peak at the zero crossings where modulation/demodulation occurs.

OFDM systems provide amongst other advantages robustness to multi-path and frequency selective fading, simple equalization and a better spectral efficiency due to enhanced modulation techniques, optimized capacity due to flexibility in time and frequency domains, an expanded coverage provided by sub-channelization and a scalable design to support extra demand giving preserved radio performance.

2.1 Sub-carriers

To achieve an OFDM system two conditions must apply:

i. In an FFT interval every subcarrier has a certain number of integer cycles.
ii. The difference between adjacent carriers is one cycle long.

This condition is satisfied by the equation 1.1 with the property expressed as a complex conjugate function.

$$\int_{t_s}^{t_s+T_s} e^{-j2\pi\frac{k}{T_s}(t-t_s)} \cdot \sum_{n=0}^{N-1} d_n e^{j2\pi\frac{n}{T_s}(t-t_s)}\,dt = \sum_{n=0}^{N-1} d_n \int_{t_s}^{t_s+T_s} e^{j2\pi\frac{n-k}{T_s}(t-t_s)}\,dt = \delta_{nk} \mid -T_g \le t \le T_s \qquad (1.1)$$

where N is the number of subcarriers, T_s is symbol time, δ is the Kronecker delta function

The adjacent sub-carrier separation shown in Fig. 3 is given by $1/_{T_s}$ and the peak values are found at $n\left[1/_{T_s}\right]$ where $n = 1,2,3\ldots L$ is an integer with a maximum L value equivalent to the maximum possible number of sub-carriers.

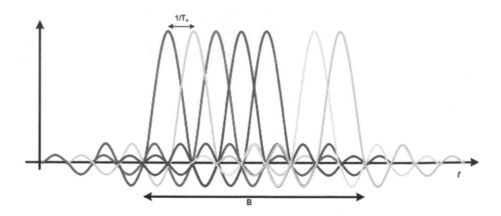

Fig. 3. OFDM Spectra of Sub-Carriers

Ideally, in a high data rate system the channel delay spread τ is much smaller than the symbol time T_s to avoid ISI. In wideband communications we require a much smaller symbol time. It's therefore, crucial to modulate the available spectrum into L sub-carriers, several discrete narrowband channels, to reduce ISI to ensure $\tau \ll T_s$ or $T_s/_L \gg \tau$. The L concurrent sub-channels are then used to send the total desired data rate ISI free, see Fig 3 [9]. Usually Inter-Carrier Interference (ICI) is easily eliminated by transmitter-receiver synchronization

2.2 Cyclic prefix

Also known as the guard interval, the Cyclic Prefix (CP) is the overhead in the time domain of an OFDM system that utilizes the delay spread due to multipath. When available spectrum is spread into several narrow-band subcarriers the symbol time increases and the opportunity to improve spectral efficiency and robustness by introducing overhead in Time Domain is possible.

By a careful estimation of the delay spread and hence a reasonable cyclic prefix T_g the ISI can be eliminated or reduced to negligible levels. The net effect of the guard interval is that all multipath effects only affects the guard interval and not the actual data symbol while the T_g remains small enough to be ignored in useful symbol time T_s [7]. The symbol duration, T is thus composed of the useful symbol time T_s and the guard interval T_g. as illustrated in Fig 4.

In attempting to create an ISI free channel the channel must appear to provide a cyclic convolution, a major property of I/FFT as we shall see in the next section. The idea of a cyclic prefix is thus integral to interference free multi-carrier wideband technology.

Consider a maximum channel delay spread, $\tau = u + 1$, if you add $T_g = u$ then we can then consider the entire bit-stream as a single OFDM symbol with L vector-lengths.

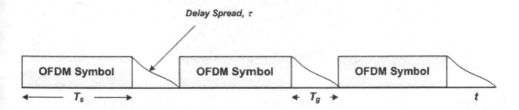

Fig. 4. OFDM Symbol

$$X = [x_1 x_2 x_3 \dots x_L] \tag{1.2}$$

$$X_{cp} = [x_{L-u} x_{L-u+1} x_{L-u+2} \dots x_{L-1} x_0 x_1 \dots x_{L-1}] = x_{cp} x_d \tag{1.3}$$

where x_{cp} is the cyclic prefix and x_d is the original data

The channel output is given by, $Y_{cp} = h \circledast X_{cp}$, where h is a length $u + 1$vector describing the impulse response of the channel during the OFDM symbol and the length of Y_{cp} is $L + 2u$ samples with one u from the previous symbol discarded and the other u discarded at the next symbol leaving only the L symbols as, originally intended, output. The idea thus to represent the signal as a circular convolution system with CP that is at least as long as the channel delay spread results in a desired channel output Y to be decomposed into a simple product of the channel frequency response H = FFT{h} and the channel frequency domain input, X = FFT{x} [9].

2.3 I/FFT

OFDM employs an efficient computational technique known as the Fast Fourier Transform (FFT) and its inverse, the Inverse First Fourier Transform (IFFT). An FFT transforms or decomposes into its frequency components while the IFFT will reverse the signal to the special domain. The FFT is a faster algorithm of the Discrete Fourier Transform (DFT) with time savings of up to a factor of $[N/logN]$.

If we consider a data sequence $X = (X_0, X_1, \dots X_n, \dots X_{N-2}, X_{N-1})$and $X_k = A_k + jB_k$ then a DFT/IDFT representation of an OFDM signal can be expressed thus, 1.4 below

$$x_n \frac{1}{N} \sum_{k=0}^{N-1} X_k e^{j2\pi \frac{kn}{N}} = \frac{1}{N} \sum_{k=0}^{N-1} X_k e^{j2\pi f_k t_n}, \; n = 0,1,2..N-1 \tag{1.4}$$

Where $f_k = {}^k/_{N\Delta t}, t_n = n\Delta t$ and Δt is an arbitrary symbol duration of the = sequence x_n

If we take the real part as $S_n = Re(x_n)$

$$= \frac{1}{N} \sum_{k=0}^{N-1} (A_k cos2\pi f_k t_n - B_k cos2\pi f_k t_n, \quad n = 0,1,2..N-1 \tag{1.5}$$

Applied to a low-pass filter we $t_n = t \; intervals, 0 \le t \le N\Delta t \; in \; equation \; 1.5 \; above$.

The time and frequency domain representations can be given as in Eq 1.6.

$$\int_{-\infty}^{\infty} x_i(t)x_j^*(t)\,dt = \begin{cases} 1, i = j \\ 0, i \neq j \end{cases} \text{ and } \int_{-\infty}^{\infty} X_i(f)X_j^*(f)\,df = \begin{cases} 1, i = j \\ 0, i \neq j \end{cases} \tag{1.6}$$

The time-domain spreading is achieved by repeating the same information in an OFDM symbol on two different sub-bands giving frequency diversity while the frequency-domain spreading is achieved by choosing conjugate symmetric inputs to the IFFT. This also exploits frequency diversity and minimizes the transmitter complexity and improves power control.

In section 2.2 we introduced cyclic prefix and the crucial role of circular convolution applied to a linear-time invariant FIR. We shall illustrate this further to help understand the I/FFT processing in an OFDM system.

Suppose we were to compute the output y[n] of a system as a circular convolution of its impulse response h[n] and the channel input x[n] [9].

$$y[n] = h[n] \circledast x[n] = x[n] \circledast h[n] \tag{1.7}$$

where $h[n] \circledast x[n] = x[n] \circledast h[n] \triangleq \sum_{k=0}^{L-1} h[k]x[n-k]_L$ with the circular function $x[n]_L = x[n \bmod L]$ is periodic with period L.

We can thus define the output as a DFT{ y[n]} in time and frequency as given in Eq. 1.8

$$DFT\{y[n]\} = DFT\{h[n] \circledast x[n]\} \text{ and } Y[m] = H[m]X[m] \tag{1.8}$$

The L point DFT is then defined by

$$DFT\{x[n]\} = X[m] \triangleq \frac{1}{\sqrt{L}}\sum_{n=0}^{L-1} x[n]e^{-j\frac{2\pi nm}{L}} \tag{1.9}$$

with the inverse, IDFT defined by

$$IDFT\{X[m]\} = x[n] \triangleq \frac{1}{\sqrt{L}}\sum_{n=0}^{L-1} X[m]e^{j\frac{2\pi nm}{L}} \tag{2.0}$$

In summary an OFDM system may be viewed as a functional block diagram shown in Fig. 5.

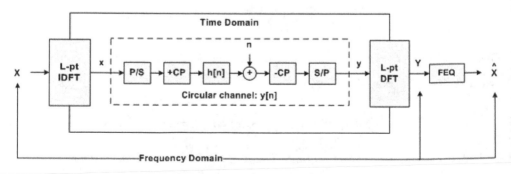

Fig. 5. OFDM Functional Block Diagram

From Fig. 5 the estimated data symbols, $\hat{X}[m] = \frac{Y[m]}{H[m]}$ while X and Y represent the L transmitted and received symbols.

i. Decompose the wideband signal of bandwidth B into L narrowband, flat-fading signals, vector x.
ii. Modulate the L subcarriers into a single wideband Radio using an IFFT operation.
iii. Attach a CP after the IFFT operation to achieve orthogonality and send serially through the wideband channel.
iv. At the receiver, discard CP, and demodulate using an FFT operation, which results in L data symbols.
v. Equalize each subcarrier using FEQ by dividing by the complex channel gain H[i] for that subcarrier [6, 7, 9].

3. OFDMA and SOFDMA

To enhance the performance of OFDM flexible extensions have been developed. These include Orthogonal Frequency Division Multiple Access (OFDMA) and Scalable OFDMA (SOFDMA). The OFDMA feature schedules a varying number of subcarriers to each subscriber depending on needs, channel conditions or both. This gives rise to the concept of flexible sub-channelization of the bandwidth. The access domain is further enhanced by multiple access technologies of FDMA and TDMA. SOFDMA on the other hand is used on the transmission end to 'scale' the channel bandwidth. By enabling the adjustment of the FFT size hence the number of carriers to the transmission channel bandwidth, SOFDMA brings scalability to OFDM. OFDMA thus incorporates features of the Code Division Multiple Access (CDMA) by combining OFDMA and allowing low-data-rate users to transmit continuously at lower power with shorter and constant delay. A similar concept applies in the frequency domain and TDMA where the resources are partitioned in the time-frequency space, and slots are assigned along the OFDM symbol subcarrier indices [8].

Certain frequency selective impacts can be minimized by spreading subcarriers of a user over the entire channel spectrum in addition S/OFDMA can be configured for Adaptive Antenna Systems (AAS) enhancing payload and coverage. OFDMA also comes with better high-performance coding techniques such as Turbo Coding and Low-Density Parity Check (LDPC), enhancing security and NLOS performance increase system gain by use of denser sub-channelization, thereby improving indoor penetration [7]. The sum effect of sub-channelization is that the link budget is greatly enhanced by allocating each subscriber one or several portions of the overall bandwidth and NOT the entire channel bandwidth. The immediate benefit of this technique is not hard to discern since the UL-DL budgets can then be easily balanced. With a good scheduling technique a fair throughput trade-off can be optimized at cell edge and overall UL data rate.

An OFDMA transmission is shown in Fig. 6 in the frequency domain, with the Pilot Subcarriers used for channel estimations, the DC subcarrier is the centre frequency, unused, the SS1/2 Data Subcarriers are the user data subcarriers scaled for subscribers 1&2 and the Guard band is to limit ISI and channel decay.

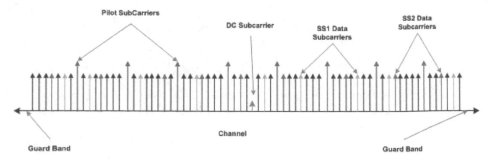

Fig. 6. OFDM Subcarriers

4. Subcarrier permutations

To create an OFDM symbol in the frequency domain, the mapping of the physical resources (subcarriers) to the logical channels (sub-channels) has to be done. This is not only aimed at assigning the right modulated signals with the right transmission blocks but also meant at reducing sub-channel sensitivity with regards to spectral fading. Permutation scheme is used to carry out this mapping. This ensures a sub-channel can use its assigned subcarrier only for a finite number of symbols and then permutated to another subcarrier. In general 802.16* suite of standards define a number permutation schemes for varied requirements ideally to introduce a robustness that minimizes interference. These schemes could be [2, 3]:

- Full or Partial sub-channel permutations.
- Distributed or Adjacent permutations.
- Uplink and Downlink permutations.

Distributed permutations use the full spectral diversity of the subcarriers for the permutation of a sub-channel while adjacent permutations assign adjacent sub carriers to a sub-channel. The distributed mode is ideal for optimizing a network towards a more robust spectral sensitivity. On the other hand the adjacent sub carriers allow faster system feedback and permutation processing thus better suited for fixed/portable devices with increased throughput [7].

4.1 Segmentation and sub-channelization

A sub-channel is a logical transmission resource of a collection of physical subcarriers. A permutation scheme defines the number and pattern for mapping the subcarriers to the sub-channels. A sub-channel is constant in time over a transmission block and maybe allocated to different connections over time. However, subcarriers of a sub-channel do not have to be adjacent. Certain factors such as size of data block to be transmitted, the modulation scheme and the coding rate may determine the amount of sub-channels allocated to a specified data block. It's important to note that a particular data region of users always uses the same burst profile. A burst profile is defined by a chosen modulation scheme, coding rate and FEC type while a data region of users refers to the contiguous set of sub-channels assigned to a user(s) in frequency and time [8,9].

An optional alternative to sub-channelization is segmentation shown in Fig. 7 that aims to divide a transmission channel into groups of sub-channels with the following properties:

- A segment consists 1 of a number of sub-channels.
- The segments share the bandwidth of the transmission channel.
- No reuse of Sub-channels of one segment and consequently the subcarriers cannot be reused either.
- Segmentation is done by interleaving the subcarriers in the Frequency Domain.
- Every segment has its own MAC instance complete with a preamble, UL MAP, DL MAP (a whole separate transmission frame).
- One subcarrier can be used only in one segment at a given time.

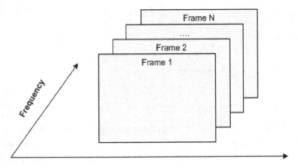

Fig. 7. Segmentation

4.2 FUSC permutation

In the Fully Used Sub-Channelization permutation shown in Fig. 8, all subcarriers are used in all the sub-channels distributed evenly across the entire frequency band. FUSC is only permutated in the Downlink (DL). The set of Pilot subcarriers, which are assigned first, is divided into two constant and two variables sets. The difference in both sets lies in the indexing of the pilot subscribers. With the variable set the index changes from one OFDM symbol to the next, while the index stays constant with the constant set. The variable sets allows for accurate estimation of channel response at the receiver especially in channels with larger delay spread or small coherence bandwidth. In cases where FUSC is implemented with transmit diversity, say of order n, then each antenna is allocated an nth each of the variable and constant sets of pilot subscribers [4, 9].

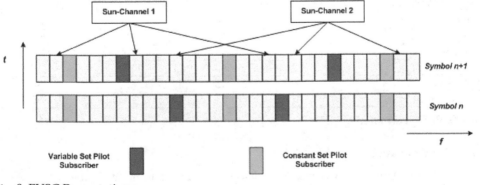

Fig. 8. FUSC Permutation

It's instructive to note that each sub-channel has a max of 48 subcarriers across all FFT sizes as shown in Table 2. The 802.16d does NOT support FUSC/PUSC and thus the 256 FFT size is not ignored in the presentation.

	128	512	1024	2048
Subcarriers per Channel	48	48	48	48
Sub-channels	2	8	16	32
Data SCa	96	384	768	1536
Pilot SCa – Constant Set	1	6	11	24
Pilot SCa – Variable Set	9	36	71	142
Left Guard SCa	11	43	87	173
Right-Guard SCa	10	42	86	172

Table 2. Parameters of FUSC Permutation

4.3 PUSC permutation

Partially Used Sub-Channelization is based on the concept of segmentation with subcarriers allocated to a segment first then to the sub-channel belonging to the dedicated segment. PUSC is similar to FUSC but with the extra advantage of permutation both in the UL and DL. The subcarriers are first subdivided into groups of 6 then clustered, save for the null subcarrier. The clusters consist of fourteen adjacent subcarriers spanned over two OFDM symbols. Permutations are thus done within groups independently of the others [12].

In the DL, each cluster's subcarriers are divided into 24 data subcarriers and 4 pilot subcarriers. The clusters are then pseudo-randomly renumbered using a scheme that redistributes the logical identity of the clusters, then divided into six groups, with the first one-sixth of the clusters belonging to group 0, and so on. A sub-channel is created using two clusters from the same group [9]. The segmentation can be done to allocate all or a subset of the six groups to a given transmitter. If this is done over sectors of a BS a better frequency reuse can be achieved.

In the UL, the subcarriers are first divided into various tiles, consisting of 4 subcarriers over three OFDM symbols. The subcarriers within a tile are divided into eight data subcarriers and four pilot subcarriers

	128	512	1024	2048
Subcarriers per Channel	14	14	14	14
Sub-channels	3	15	30	60
Data SCa	72	360	720	1140
Pilot SCa	12	60	120	240
Left Guard SCa	22	46	92	184
Right-Guard SCa	21	45	91	183

Table 3. PUSC Permutation - DL

4.4 AMC permutation

Advanced Modulation and Coding uses adjacent subcarriers to build a sub-channel. As in the TUSC scheme, it is mainly utilized in the AAS networks. In spite of some loss of frequency diversity, exploitation of multiuser diversity is easier and robust. Multiuser diversity provides significant improvement in overall system capacity and throughput, since a sub-channel at any given time is allocated to the user with the highest SNR/capacity in that sub-channel [12].

The wireless channel is dynamic and diverse users get allocated on the sub-channel at different instants in time uncorrelated channel conditions. In AMC permutation, nine adjacent subcarriers with eight data subcarriers and one pilot subcarrier are used to form a bin, as shown in Fig. 9. An AMC sub-channel consists of six contiguous bins from within the same band where four adjacent bins in the frequency domain constitute a band. An AMC sub-channel thus consists of one bin over six consecutive symbols, two consecutive bins over three consecutive symbols or three consecutive bins over two consecutive symbols [9, 10].

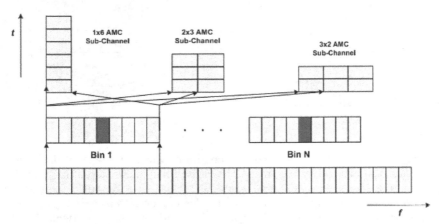

Fig. 9. AMC Permutation

4.5 TUSC permutation

The Tile Usage of Sub-channels is a downlink subcarrier permutation mode that is identical to the uplink PUSC. If closed loop advanced antenna systems (AAS) are to be used with the PUSC mode, explicit feedback of the channel state information (CSI) from the MS to the BS would be required even in the case of TDD, since the UL and DL allocations are not symmetric, and channel reciprocity cannot be used. TUSC allows for a DL allocation that is symmetric to the UL PUSC, thus taking advantage of UL and DL allocation symmetry and eliminating the requirement for explicit CSI feedback in the case of closed-loop AAS for TDD systems. Refer to TTD frame structure shown in Fig. 10.

5. Slot and frame structure

The MAC layer allocates the time/frequency resources to various users in units of *slots*, the smallest quanta of PHY layer resource that can be allocated to a single user in the time/frequency domain. The size of a slot is dependent on the subcarrier permutation mode as discussed in section 4.

- FUSC: Each slot is 48 subcarriers by one OFDM symbol.
- Downlink PUSC: Each slot is 24 subcarriers by two OFDM symbols.

- Uplink PUSC and TUSC: Each slot is 16 subcarriers by three OFDM symbols.
- Band AMC: Each slot is 8, 16, or 24 subcarriers by 6, 3, or 2 OFDM symbols.

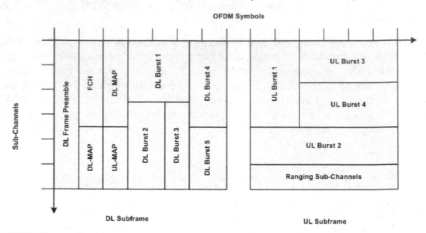

Fig. 10. TDD Frame Structure

6. Channel coding

In the 802.16e definition channel coding constitutes the sum of steps including data randomization, channel coding, rate matching, Hybrid Automatic Repeat reQuest (HARQ), and interleaving. At the beginning of each FEC block is a modulo-2 shift-register of maximum-length whose output is used to randomize the data. This randomization is purely for data integrity by providing PHY encryption and avoiding accidental decode by unintended receivers. In cases where HARQ is used the initial seed of the shift-register is kept the same over the period to allow for joint FEC decoding over several transmissions.

Channel coding is performed on every FEC block, which is n integer long sub-channels and whose maximum depends on channel coding scheme and the modulation constellation. Should the required number of FEC block sub-channels exceed this maximum, then segmentation is done to produce multiple FEC sub-blocks. Encoding and rate matching is done separately for these sub-blocks and then concatenated sequentially [10].

6.1 Convolution coding

The default channel coding scheme for WiMAX channels a convolutional encoder; based on binary non-recursive convolutional coding (CC). It uses a constituent encoder with a constraint length 7 and a native code rate $^1/_2$. The convolutional encoder transforms an m-bit symbol into an n-bit symbol, where m/n is the code rate $\{n \geq m\}$. The transformation is a function of the last k information symbols, where k is the constraint length of the code [5].

The Turbo encoder is used to encode the output of the data randomizer. A padding byte 0x00 at the end of the OFDM mode of each FEC block is used to initialize the encoder to the 0 state. In OFDMA tailbiting is used to initialize the encoder by using 6 bits from the end of the data block appended at the beginning, to be used as flush bits. These appended bits flush out the bits left in the encoder by the previous FEC block. The first 12 parity bits that are generated by the convolutional encoder which depend on the 6 bits left in the encoder by the previous FEC block are discarded [9].

6.2 Turbo coding

Turbo codes are high-performance error correction codes used to achieve maximal information transfer over a bandwidth-limited noise prone communication.

WiMAX uses duo-binary turbo codes with a constituent recursive encoder of constraint length of 4. In duo-binary turbo codes, two consecutive bits from the un-coded bit sequence are sent to the encoder simultaneously. The duo-binary convolution encoder has two generating polynomials, 1+D2+D3 and 1+D3 for two parity bits with four possible state transitions.

6.3 Block turbo and LDPC coding

These are some of the optional channel coding schemes for WiMAX. The block turbo codes consist of two binary extended Hamming codes that are applied on natural and interleaved information bit sequences, respectively. The LDPC code, is based on a set of one or more fundamental LDPC codes, each of the fundamental codes is a systematic linear block code that can accommodate various code rates and packet sizes. The LDPC code can flexibly support various block sizes for each code rate through the use of an expansion factor [9, 12].

6.4 HARQ

Hybrid Automatic Repeat reQuest can be implemented in type I or II, commonly referred to as chase combining and incremental redundancy respectively. In Chase combining the receiver uses the current and all the previous transmissions of data block for puncturing and hopefully the right information is decoded or the HARQ timer runs out. However, with incremental redundancy the receiver uses different version of the redundancy block leading to lower BER and BLER.

7. Symbol mapping and structure

When a symbol is mapped, the sequence of binary bits is converted to a sequence of complex valued symbols. The mandatory constellations are QPSK and 16 QAM, with an optional 64QAM constellation also defined in the IEEE 802.16e standard. Assuming all symbols are equo-probable each modulation constellation is scaled by a factor ς such that the average transmitted power is unity. Where ς is given by $\sqrt{1/2}, \sqrt{1/10}, \sqrt{1/42}$ for QPSK, 16QAm and 64QAM modulations shown in Fig. 11. If convolution coding is applied for PHY encryption then we have an output given by

$$S_k = 2(1/2 - w_k)S_k$$

By scaling the preamble and midamble symbols by a factor $2\sqrt{2}$ we we amplify the power by a factor 8 and allow for more accurate synchronization and various parameter estimations, such as channel response and noise variance.

A high-data-rate sequence of symbols can be split into multiple parallel low-data rate-sequences, each of which is used to modulate an orthogonal tone, or subcarrier. The resulting baseband signal, which is an ensemble of the signals in all the subcarriers, can be represented as

$$x(t) = \sum_{i=0}^{L-1} s[i] e^{-j2\pi(\Delta f + iB_\varsigma)t} \quad 0 \le t \le T$$

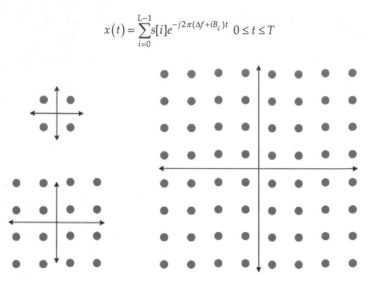

Fig. 11. QPSK, 16QAM and 64QAM Modulation Constellations

8. References

[1] IEEE WMAN 802.16, http://ieee802.org/16/
[2] IEEE. Standard 802.16-2004, Part 16: Air interface for fixed broadband wireless access systems, June 2004.
[3] IEEE. Standard 802.16-2005, Part 16: Air interface for fixed and mobile broadband wireless access systems, December 2005.
[4] www.wimaxforum.org
[5] SR Telecom, "Wimax Capacity", 2006
[6] S. H. Han and J. H. Lee. "An overview of peak-to-average power ratio reduction techniques for multicarrier transmission." IEEE Wireless Communications, 12(2):56–65, 2005
[7] Alactel Feature Description, "WiMAX RAN System Introduction", 2 Ed.
[8] Hujun Yin and Siavash Alamouti (August 2007). "OFDMA: A Broadband Wireless Access Technology". IEEE Sarnoff Symposium, 2006
[9] J. G. Andrews, A. Ghosh and R. Muhamed, "Fundamentals of WiMAX: Understanding Broadband Wireless Networking," Prentice-Hall, June 2007
[10] Hassan Yaghoobi, "Scalable OFDMA Physical Layer in IEEE 802.16 WirelessMAN", Intel Technology Journal, vol 8, Issue 3, Aug 2004
[11] Amimo-Rayolla, Anish Kurien, Chatelain Damien and Odhiambo Marcel, "Wireless Broadband: Comparative Analysis of HSDPA vs. WiMAX", SATNAC, 2007.
[12] Marcos D. Katz, Frank H. P. Fitzek, "WiMAX evolution: Emerging Technologies and Applications", John Wiley and Sons, 2009

Permissions

The contributors of this book come from diverse backgrounds, making this book a truly international effort. This book will bring forth new frontiers with its revolutionizing research information and detailed analysis of the nascent developments around the world.

We would like to thank C. Palanisamy, for lending his expertise to make the book truly unique. He has played a crucial role in the development of this book. Without his invaluable contribution this book wouldn't have been possible. He has made vital efforts to compile up to date information on the varied aspects of this subject to make this book a valuable addition to the collection of many professionals and students.

This book was conceptualized with the vision of imparting up-to-date information and advanced data in this field. To ensure the same, a matchless editorial board was set up. Every individual on the board went through rigorous rounds of assessment to prove their worth. After which they invested a large part of their time researching and compiling the most relevant data for our readers. Conferences and sessions were held from time to time between the editorial board and the contributing authors to present the data in the most comprehensible form. The editorial team has worked tirelessly to provide valuable and valid information to help people across the globe.

Every chapter published in this book has been scrutinized by our experts. Their significance has been extensively debated. The topics covered herein carry significant findings which will fuel the growth of the discipline. They may even be implemented as practical applications or may be referred to as a beginning point for another development. Chapters in this book were first published by InTech; hereby published with permission under the Creative Commons Attribution License or equivalent.

The editorial board has been involved in producing this book since its inception. They have spent rigorous hours researching and exploring the diverse topics which have resulted in the successful publishing of this book. They have passed on their knowledge of decades through this book. To expedite this challenging task, the publisher supported the team at every step. A small team of assistant editors was also appointed to further simplify the editing procedure and attain best results for the readers.

Our editorial team has been hand-picked from every corner of the world. Their multi-ethnicity adds dynamic inputs to the discussions which result in innovative

outcomes. These outcomes are then further discussed with the researchers and contributors who give their valuable feedback and opinion regarding the same. The feedback is then collaborated with the researches and they are edited in a comprehensive manner to aid the understanding of the subject.

Apart from the editorial board, the designing team has also invested a significant amount of their time in understanding the subject and creating the most relevant covers. They scrutinized every image to scout for the most suitable representation of the subject and create an appropriate cover for the book.

The publishing team has been involved in this book since its early stages. They were actively engaged in every process, be it collecting the data, connecting with the contributors or procuring relevant information. The team has been an ardent support to the editorial, designing and production team. Their endless efforts to recruit the best for this project, has resulted in the accomplishment of this book. They are a veteran in the field of academics and their pool of knowledge is as vast as their experience in printing. Their expertise and guidance has proved useful at every step. Their uncompromising quality standards have made this book an exceptional effort. Their encouragement from time to time has been an inspiration for everyone.

The publisher and the editorial board hope that this book will prove to be a valuable piece of knowledge for researchers, students, practitioners and scholars across the globe.